MW01008889

DESIGNING WITH
LIGHT
The Creative Touch

Carol Soucek King, M.F.A., Ph.D.

Foreword by Stanley Abercrombie, FAIA

PBC INTERNATIONAL, INC.

Distributor to the book trade in the United States and Canada
Rizzoli International Publications Inc.
through St. Martin's Press
175 Fifth Avenue
New York, NY 10010

Distributor to the art trade in the United States and Canada
PBC International, Inc.
One School Street
Glen Cove, NY 11542

Distributor throughout the rest of the world
Hearst Books International
1350 Avenue of the Americas
New York, NY 10019

Library of Congress Cataloging-in-Publication Data
King, Carol Soucek.
Designing with light : the creative touch / Carol Soucek King.
p. cm.
Includes index.
ISBN 0-86636-582-6 (hardcover). — ISBN 0-86636-583-4 (pbk.)
1. Lighting, Architectural and decorative. 2. Interior decoration. I. Title.
NK2115.5.L5K56 1997
747'.92—dc21 97-22718
CIP

CAVEAT—Information in this text is believed accurate, and will pose no problem for the student or casual reader. However, the author was often constrained by information contained in signed release forms, information that could have been in error or not included at all. Any misinformation (or lack of information) is the result of failure in these attestations. The author has done whatever is possible to insure accuracy.

10 9 8 7 6 5 4 3 2 1

Printed in Hong Kong

To Creativity at Home...
and Being at Home with Creativity!

contents

BEN SHAHN

Sorry to repeat myself, but in the last sentence of a book titled *A Philosophy of Interior Design,* I wrote that interiors constitute "our most personal art." Carol Soucek King, I'm glad to see, seems to share the same view, for in her admirable series of books Dr. King focuses on just those aspects of interior design that make it personal.

The grand concept is not to be neglected, of course. Like any other art, interior design depends for its success on the encompassing vision that relates its many elements in a meaningful whole. But such vision, in interiors, becomes manifest and comprehensible through the myriad details with which we are in intimate contact: the feel of a drawer-pull, the profile of a cornice, the polish and grain of wood, the "hand" of fabric.

This contact involves all our senses. We see our interiors, certainly, but we also smell the materials in them, we hear their acoustic properties, we brush up against their walls, step on their floors, open their case goods, sit on their chairs. More than any other, interior design is the art we use. In that sense, it is not only our most personal art, but also the one most responsible for our well-being. In the context of increasingly brutalized urban enviroments, this is increasingly true and increasingly important. Interior design is often our refuge.

It is therefore a very welcome prospect that Dr. King is turning her experienced editorial eye to the details and materials on which the art of interior design depends. I'm sure we will all benefit from her discoveries.

~Stanley Abercrombie, FAIA
Chief Editor, Interior Design

FOREWORD

PREFACE

Welcome to DESIGNING WITH LIGHT, the fifth book in "The Creative Touch" series published by PBC International, Inc. Similar to *Designing with Tile, Stone & Brick, Designing with Wood, Designing with Glass* and *Designing with Fabric,* this volume is also devoted to one aspect of residential interiors. Yet it is distinct, too, for light is an element capable of visually sculpting and elevating all others.

©Anthony Peres

"Because light alone among the designer's tools is constantly changing throughout the day, with one light source capable of expressing so many different moods, it is the home's fourth dimension," says Holland's lighting designer Aleksandar Rublek. To Mexico's architect Ricardo Legorreta, light is above all a powerful emotional force that, despite its physical and technological roots, "belongs more to the heart and to the spirit." And to Japan's lighting designer Motoko Ishii, "Light is life!"

Clearly, light is both highly scientific and deeply psychological. It should enhance a home's function and at the same time reveal its poetry. What do the people who live there really want — showcase or understatement? Mood or subtle ambience? As much natural light as possible — or round-the-clock drama? On the following pages, these questions and more have been addressed with great sensitivity by leading architects and designers around the world. Their creative solutions serve as a testament both to lighting's advanced technology — and to its soul. Enjoy!

~**Carol Soucek King,** M.F.A., Ph.D.

Light and spirituality go together. Light and architecture, windows, materials, textures and colors. During hours, days and seasons it changes space and is a fundamental tool for shaping our emotional response. Light, both natural and artificial, cannot be ignored nor used with a technical mind. Light belongs to the heart and to the spirit. Light attracts people, it shows the way, and when we see light in the distance, we follow it. We seek light. Light, both natural and artificial, can be sculpture.

Louis Kahn and others have said that form appears when light falls upon it, and of course I agree. But the opposite is true as well. Floors and walls, ceilings and recesses allow us to sculpt with light and to paint with light and, in this way, to shape the ambience of space. Light brings out the character of traditional buildings, which in turn inspires my own efforts in designing new buildings.

~Ricardo Legorreta

Photography by Lourdes Legorreta

Introduction

Traditions
Aglow

Ranch House Comfort

Northern Louisiana

Photography by ©1996 Ira Montgomery

In remodeling this traditional Southern ranch house, **Mil Bodron** combined a feeling of classicism and modernism within a traditional envelope. Following suit, **Barbara Bouyea's** lighting design enhanced the architecture without being blatant. The artwork and architecture were key and no other supportive design elements — such as lighting — could show.

"In a ranch-style house, ceilings are normally low and there is a sense of small volumes," says Bouyea. "The lighting was used to enhance and enlarge the feeling of volume in each space.

"I believe the best lighting in a residence is where people are comfortable and look great — through a tone that enhances the skin, and where lighting is flattering and doesn't create harsh shadows. Yet also important is the illumination of artwork and accessories, of the ceiling, walls and floors that define a space, and of the textures, forms and hues that give it character."

To arrive at such a lighting plan, achieving a balance between ambient and accent lighting is critical, and so is the correct dimming system, which can create a variety of moods in each space. Bouyea has made sure that all spaces vary in function, and so does the lighting.

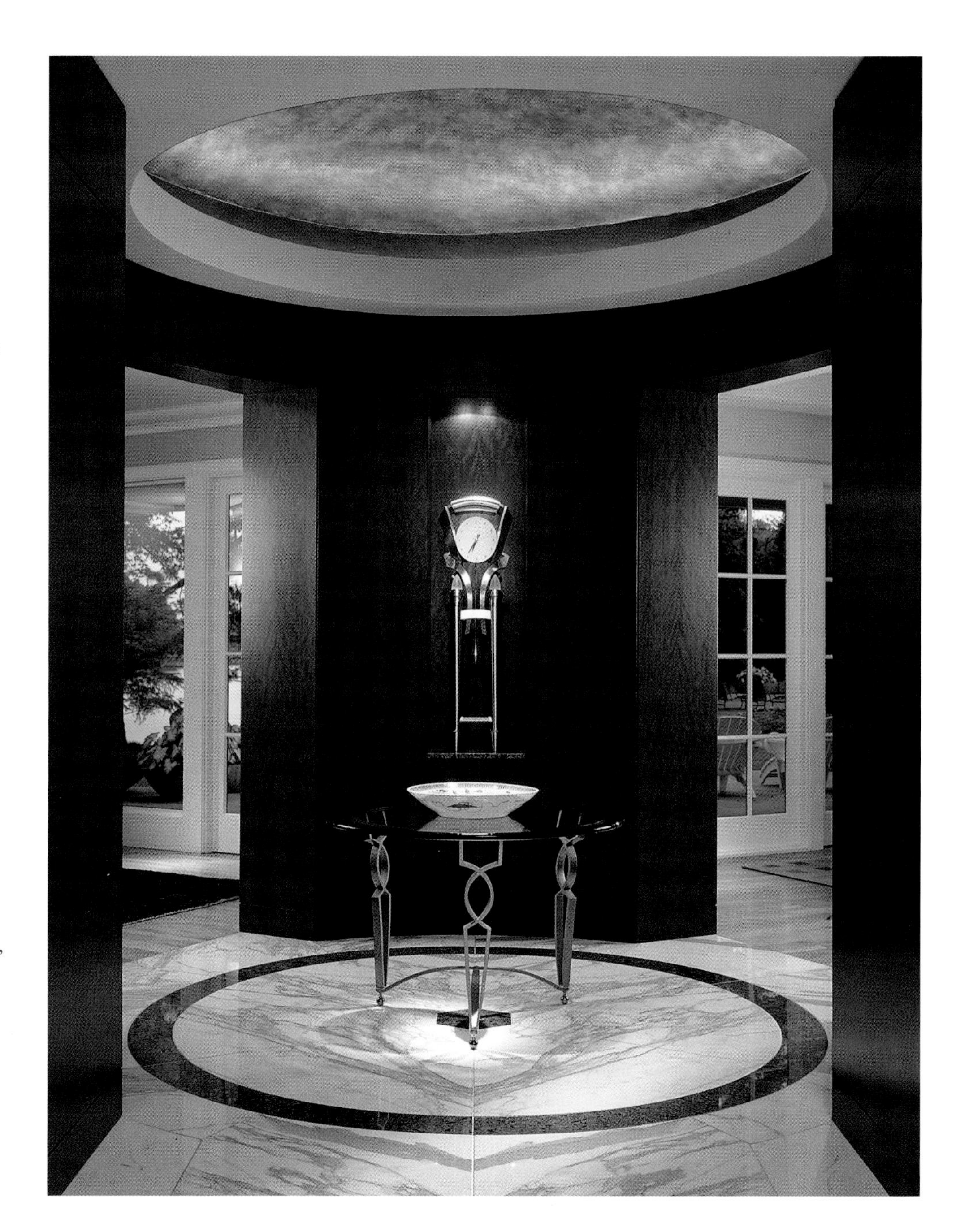

above At the entry, spaces and volumes are defined by lighting located at the platinum domed ceiling. Accent lighting highlights the niche and clock.

opposite In the living room, accent lighting for artwork and accessories creates dramatic focal points.

below In the dining room, lighting accents artwork and accessories. Cove lighting over the drapery gives play to the window curve and the fabric's shimmery quality.

bottom General ambient lighting is provided by wall sconces and floor lamps in the living room.

right In the kitchen, accent lighting highlights artwork, while task lighting illuminates work areas as well as the island and breakfast table. Lighting in the niche above the built-in cabinets leads the eye upward, and lighting in the skylight provides moonlight patterns after dusk.

above Due to structural challenges, the master bedroom's bird's-eye maple walls are illuminated by a wall of miniature downlights. At the seating area, floor lamps provide light for reading, while an accent light highlights a floral arrangement.

opposite above Coves were created not only to illuminate marble and glass walls, but also to distinguish spatial elements in the master bathroom. A pattern of miniature downlights helps define the tub area.

opposite below The powder room's glass sink is illuminated from below, defining the sink within the dark granite counter. At the mirror, wall sconces provide illumination that is highly functional, yet soft and flattering.

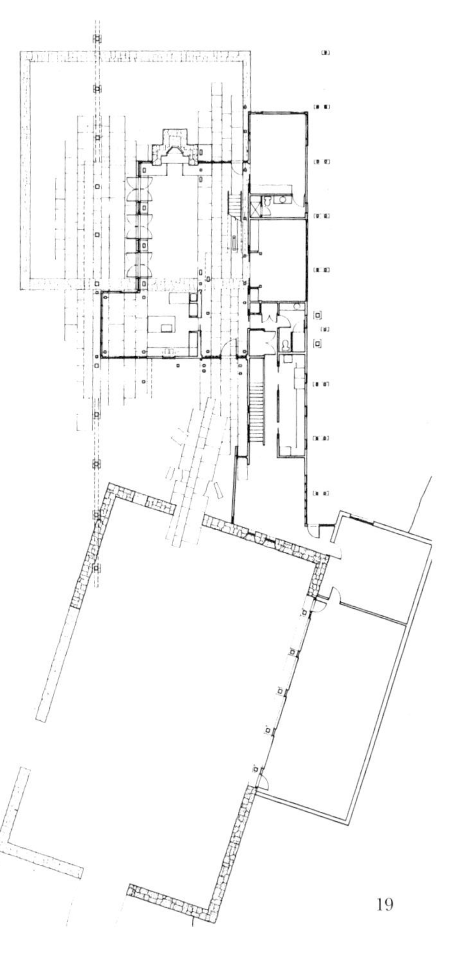

High Tech Hacienda

Los Angeles, California

Photography by © David Glomb

This home's style may be yesterday's Spanish Colonial, yet its integration of high technology makes it state-of-the-art.

Weathered antique doors, floating beams, arched doorways and a tiled roof give the impression that this house has been around for a century. In fact, however, they conceal ultramodern lighting by **Julia Rezek**.

Her lighting brings out the rich textures that architect **J. Scott Carter** and interior designer **Dennis Hague** brought to the space, especially focusing on the floating beams and exposed wood ceiling, the tile and wood flooring, the textural cabinetry and furniture, and the clients' contemporary art collection.

The clients wanted the lighting to be as understated as possible, yet feature the artfulness of their home. They also wanted the lighting to be simple to operate. The answer revealed itself in many different forms — four different lighting applications in the living room, accent lights in the dining room, a balance of task and ambient lights in the kitchen, and light expanding techniques in the master bathroom.

The lighting control system was designed to be multipurpose and user-friendly. For maximum flexibility, Rezek opted for a multi-scene preset dimmer panel so that all lighting zones within the space could be accessed with the touch of a single button.

top Decorative fixtures and candles contribute to the sense of hospitable warmth and, especially in rooms with soaring ceilings, create pools of light at eye level.

above A slotted aperture located in the dining room's flat ceiling is fitted with *MR 16* adjustable accent lighting.

right Concealed above the floating beams are linear incandescent strip lights that softly wash the exposed wood and tongue-and-groove sloped ceiling, fitted with adjustable low-voltage *MR 16* accents.

above To provide enough light in the kitchen without resorting to the track lighting often used when skylights are located over task areas, Rezek placed a low-voltage system in the underside of the skylight mullions. Lighting concealed on the bottom side of the upper cabinets is provided by a lensed fixture with fluorescent lamps.

right In the family room, recessed MR 16 accent lighting adjacent to the walls emphasizes works of art. The reflection of light off the walls eliminates the need for central downlights.

above In the master bedroom, recessed *MR 16* adjustable accent lighting is positioned in a slotted aperture at the angled ceilings over bed and seating areas. A weighted uplight at the base of the planter casts exciting patterns on the ceiling. The decorative bedside fixture was designed by Dennis Hague.

right Minimal depth over the sinks in the master bathroom led to the selection of recessed 35-watt *MR 16* lamps, with incandescent linear strip light flanking each mirror to provide fill light. *MR 16* spotlights delineate tile over the tub, and a recessed, waterproof fluorescent fixture provides a day-like glow in the shower.

Neoclassic Sophistication

Metropolitan New York

Photography by Peter Vitale

above Recessed halogen lighting, strategically located in the ceiling, illuminates the mirror and artwork. Strip lights highlight the staircase landing.

left Custom soffits allow for uplighting and downlighting while concealing the source.

A sense of reserve was in order for the lighting of this neoclassic residential statement created by **Michael Wolk**. Developed for a husband and wife who like to entertain formally, it needed a lighting plan in keeping with the traditional, elegant approach taken by Wolk in designing the interior architectural details as well as the furnishings.

Indirect uplighting hidden in soffits, and antique chandeliers and table lamps visually articulate the collection of rich fabrics, classical columns and moldings with warm, soft, incandescent light. The focal point of the urbane living room is an inviting two-sided sofa. The dining room is equally refined, with a streamlined buffet delicately balanced on massive curving capitals. In the entertainment room, the custom-designed media unit, book shelves and fireplace create a casual, yet well-bred atmosphere. Other grace notes include a serpentine staircase, the designer's contemporary interpretation of a sleigh bed, and a translucent color palette of ivory, vanilla and gold. By keeping the mood of the lighting in the same refined and traditional vein, Wolk has put the final touch on this cultivated, harmonious environment.

At a time when so many lighting effects visually shout "high-tech," such understated elegance seems almost innovative.

above & right In the dining room, an antique chandelier is the focal point as well as primary light source. Perimeter recessed lighting defines the buffet and adds ambient illumination.

below In the breakfast room, the skylight is fitted with indirect fluorescent lighting. A recessed halogen strip light in the mirror-backed niche gives the area additional glow.

above Two antique table lamps, separate in size and style, add to the eclectic sophistication of the master bedroom.
left In the family room, recessed ceiling lights provide general illumination and direct lighting at shelves.
below Color-true halogen lighting illuminates the sparkling white marble of the bathroom.

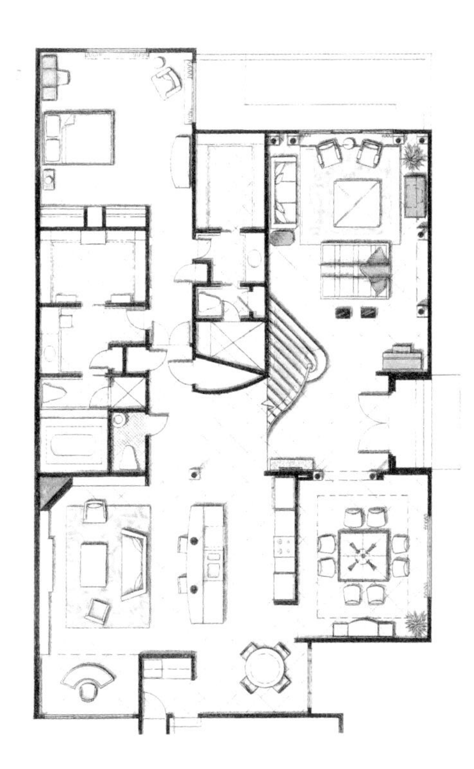

Cottage Cozy

Carmel, California

Photography by ©1994 Douglas A. Salin and Craig Mole Photography

When Ann LaCroix moved from a spacious house into a cottage, she asked architect **Wm. David Martin** to remodel and enlarge to the 1600-square-foot maximum allowed on this 40-x-40-foot lot in picturesque Carmel, California. She wanted as much light-filled interior space as possible.

Martin's answer was to partially remove the flat eight-foot ceilings and to re-frame the roof to maximize interior volumes with open beam vaults. Circulation from the existing living room to the new dining/family room was improved by creating a hallway screened from the bedroom by sliding shoji panels. A 42-foot ridge-long skylight was used to reinforce this unusual circulation requirement.

For lighting designer **Linda Ferry**, a principal challenge was to emphasize the common ceiling structure and the long skylight between the living and dining rooms, thus unifying the unique circulation pattern and creating the desired sense of space.

Says Ferry, who also worked on this project with interior designer **John Schneider** and kitchen designer **Sheron Bailey** to light task areas as well as LaCroix's Asian and Southeast Asian-inspired collection, "If done correctly, lighting reveals the composition of each design."

above A 42-foot-long skylight illuminated by a low-voltage light strip highlights the corridor between living and dining areas. Backlighting sets niches and Balinese artifacts in relief. Ambient light enhances the use of bamboo and limestone.

right In the living room, adjustable recessed low-voltage lighting by Capri emphasizes the interesting mix of Asian, Balinese and other Southeast Asian artifacts. Uplighting turns a palm into a shadow play across the ceiling, while table lamps contribute intimate, ambient light.

above Recessed downlights illuminate task areas in the kitchen. A light strip behind a reveal at the top of the cabinet washes the ceiling with light. Shoji-screen room dividers and cabinet doors define the spaces without blocking ambient light.

left Uplighting along the skylight emphasizes the structure and volume of the space. The shoji screens are illuminated from behind to increase the sense of dimension. Downlights on the dining table and buffet area, as well as a light strip under the countertop, provide warmth as well as function.

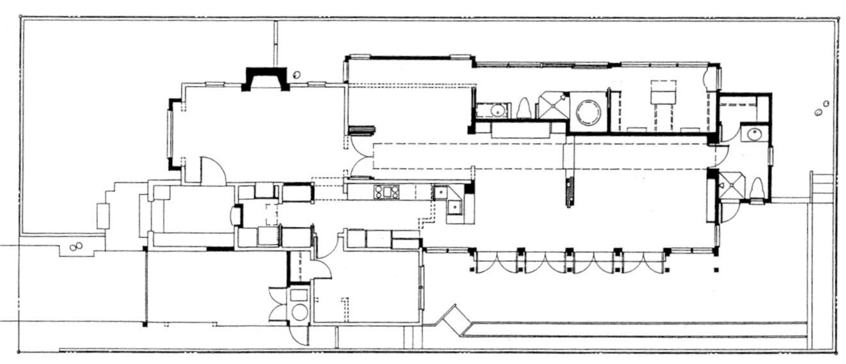

Light Music

Kansas City, Missouri

Photography by ©Michael Spillers 1997, Kansas City

above The sunroom adjacent to the living room is illuminated with a soft, incandescent ambient light. This more relaxed atmosphere complements the drama of the living room.

left In the living room, low-voltage adjustable accent fixtures are located at key activity areas and focal points. The fixtures are grouped in curving circulation patterns. In contrast, a linear row of black Alzak downlights is located along the primary window walls, washing the drapery with light — an ideal backdrop for musical performances.

Purchased by a theatrical couple—one a composer, the other a singer who love to entertain at home—this early twentieth-century grande dame was given a total face-lift to make her better suited to their lifestyle. Lighting being almost as important to them as music, they called on **Bruce Yarnell** and **Derek Porter** at the beginning of the reconstruction process so that the lighting could evolve in concert with the interior's other aspects.

Working closely with the electrical contractor in siting fixture mounting locations, they installed primarily directional halogen lighting with minimal ambient light in order to dramatically illuminate the main entertaining areas. Another important feature of such recessed architectural products is that their hardware is inconspicuous and thus respects the home's traditional style.

A highly sensitive aspect of the lighting plan is the manner in which the designers located the fixtures in the public spaces. Whereas transitional and secondary/private areas utilize a traditional, linear arrangement, ceiling fixtures are installed in serpentine curves in the living room, dining room and kitchen. Instead of being in alignment with the walls, these patterns offer additional grace notes, reflective of the way in which these rooms are used and enjoyed by groups of people.

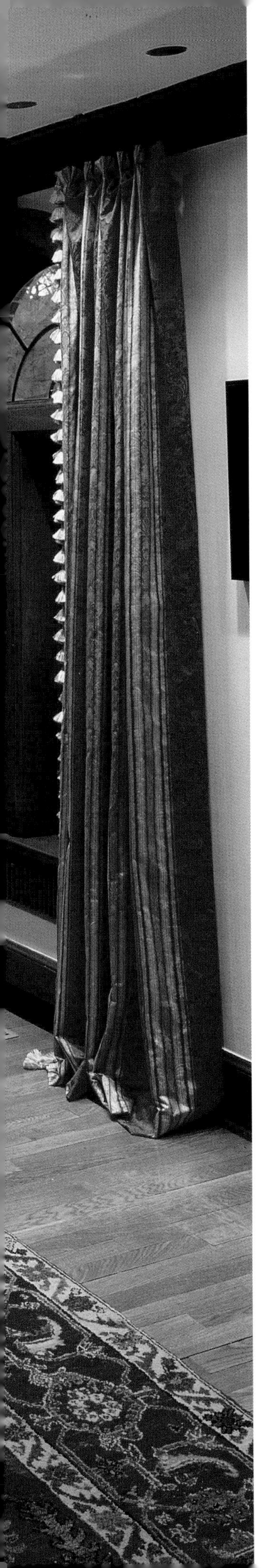

above Surface ceiling-mounted halogen fixtures produce the even ambient light required in the kitchen while also providing a decorative element. Added task illumination at work surfaces is achieved by low-voltage strip lights mounted under upper cabinets and a linear row of black Alzak downlights above the sink.

left In the dining room as in the living room, recessed low-voltage adjustable accent lights with halogen *MR 16* lamps and pinhole trim illuminate focal points, while recessed downlights wash the drapery at the front window wall with light.

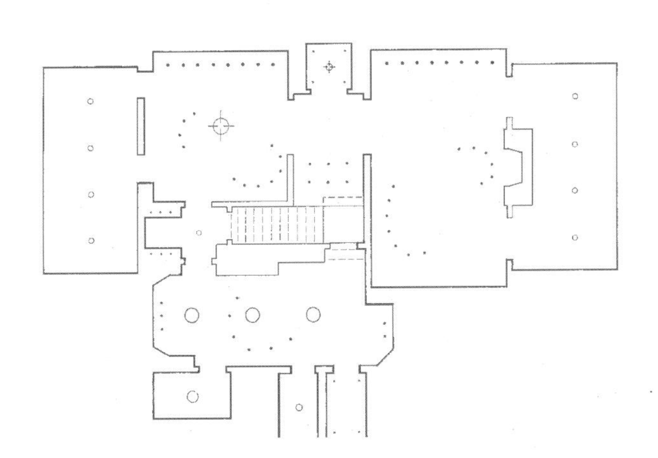

Welcoming Warmth

Georgetown, Washington, D.C.

Photography by Mick Hales

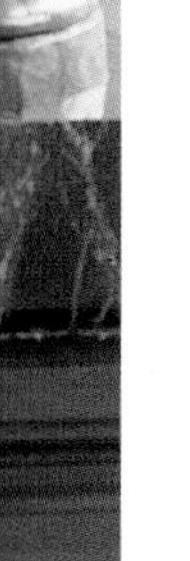

left This passage area adjacent to the dining alcove is naturally illuminated by clerestory windows that are fitted with light filtering, handwoven, Japanese grass shades. Spotlights are used only to highlight specific focal points, such as the bronze eagle and vignette.

top A traverse curtain of translucent wool, rather than heavy fabric, is used at the window wall to filter the harsh natural light. The low ceilings did not include provisions for recessed lighting; instead, fine French and Italian portable lamps with handmade silk shades cast a flattering light.

above A window-less anteroom far from the window wall is brightened by the addition of a cove ceiling fitted with uplights.

Charm is often the result of imperfections overcome. Such is the case of this Georgetown condominium in which the challenges proved to be godsends.

The building's concrete slab construction, a curved window offering no view or privacy, and an intensely bright clerestory window were transmuted by **Rita St. Clair** and her associate **Ted L. Pearson** into a welcoming environment for entertaining without a trace of its previous harshness.

To diffuse the sunlight and create privacy at the curved glass wall, they used translucent curtains, thereby controlling the illumination without cancelling it as heavier drapery would have done. They tamed the clerestory window in the dining area with natural grass shades from Japan and a tinted coating that cuts ultraviolet rays. The few ceiling junction boxes were fitted with discreet, adjustable low-voltage spotlights.

Flattering effects were created with table lamps, wall sconces and a few ceiling-mounted, low-voltage spotlights. Tinted lamps of medium wattage and lampshades in natural colors were used for a soft effect with a few specific highlights. The designers did not use track lights, as they felt the lines would conflict with the curvature of the window wall.

"Besides," says St. Clair, "a residential look was paramount, so track lighting or any other theatrical effect was not allowed."

left The anteroom ceiling is painted a golden cream color that glows when illuminated. Even though the ceiling's perimeter has been lowered, the uplighted, higher coffer seems boundless. Mirrors add to the mystery of the space and double the light effect.

below In the master bedroom, identical window treatments lend symmetry to the uneven windows. Pairs of wall sconces by Chapman provide flattering illumination at each of the wall-hung dressers.

opposite The dining alcove nestled opposite the window wall is glazed in the same deep cranberry as the anteroom. The Italian sconces give a flattering light to diners while minimal spotlighting dramatizes the centerpiece and the charger displayed on the built-in server.

American Moderne Update

Southeastern Florida

Photography by Carlos Domenech

above On top of the grand scissor staircase, a chandelier by Ingo Maurer serves as a sculptural element. A mirror mounted underneath reflects the chandelier's halogen-illuminated, gold-leafed planes.

left Two-story-high glass walls provide abundant natural light but at night, the dramatic pools of illumination cast by the antique torchieres midway up the soaring space are a necessity. Wolk found the torchieres at Miami's annual Modernist show.

Glorious as large rooms and high ceilings may be, they do not necessarily go hand-in-hand with warmth. When designing the interiors of this 14,000-square-foot waterfront home, **Michael Wolk** used lighting to make its huge rooms feel more residential in scale.

"During the day, the spaces are infused with Florida's tropical light and artificial lighting becomes secondary," says Wolk. "But at night, these grand volumes felt like a hotel, especially since all the lighting was located in the ceilings, some of which are thirty-five feet high. We had to make them less intimidating, and the use of some beautiful decorative lamps and fixtures brought the lighting down to floor level to create a sense of intimacy."

In addition, all of the furnishings were designed to reflect the owners' interest in early American Modernist influences. Wolk discarded some of the 1984 home's existing elements that dated it or that seemed more suitable for a public space, such as the bar's disco lighting and the entry's brass-and-glass railing. Wolk also considered the golden oak finish of the extensive woodwork throughout the house to be passé and spent the first three months of the project refinishing the moldings, windows and doors in deeper cherry and mahogany tones.

above Neon lizard sculptures light the hallway to the poolroom, its table illuminated by a hanging adjustable pendant fixture with halogen lamp.

left In the bar, the perimeter incandescent lighting and halogen downlights are augmented by a pendant lamp over the game table, a Modernist-style torchiere, and a table lamp.

below Perimeter lighting is provided by incandescent downlights located in the second-story and first-floor soffits. The table lamp, one of many antique lighting fixtures acquired for the home, reflects the owners' interest in birds.

above In the soffits, incandescent downlights wrap the perimeter and illuminate the home's extensive, refinished wood paneling.

right In the dining room, the custom-designed buffet is highlighted by recessed low-voltage strips. The antique vases are illuminated by halogen fixtures.

below In the powder room, an antique pendant fixture provides soft uplighting as well as downlighting. The wall sconces were designed by architect Michael Graves.

Cher-ished Properties

Malibu and Bel Air, California

Photography by Ron Wilson

A safe haven — the feeling that each room is wrapping its arms around her — is what the actress Cher wants most from a home and why she asked **Ron Wilson** to design eighteen of them over the past twenty-five years. He understands her need to blend comfort with her seemingly contradictory yearning for precision, elegance, and mood lighting.

Even though her home is located in Malibu, where daylighting and exterior views mean everything, Cher's interest is focused on the spell cast by the interior itself.

"The effect she prefers is for no artificial source of light to be apparent," says Wilson. "In fact, if it is practical, she prefers the entire interior to be candlelit."

The variety of Wilson's projects demands design versatility. As opposed to Cher's interest in mood lighting, Wilson's other clientele often indicate a preference for traditional lighting. Built in the early 1930s by architect to the rich and famous, Wallace Neff, the French Normandy-style Cohen residence (located in Bel Air) reflects his typically European approach to light and darkness. Neff could often be heard expounding on the importance of chiaroscuro—the treatment of light and shade to produce the illusion of depth. Wilson emphasized this idea in order to remain true to the original architecture's bones.

A mysterious elegance is achieved by the lighting design of Cher's home, which also reflects her need for privacy. By day, the abundance of natural light in the dining room is refracted from the exterior. Over twenty beeswax candles in the hanging fixture and on the table provide illumination at night. In the family room, light at the ceiling and introduced from floor and table lamps subtly highlights the heavily textured walls and their various natural tones. Light is reflected into the center of the room, providing warmth without resorting to color and allowing Cher herself and guests to provide the drama.

Candlelight enhances the drama of the foyer which leads to the dining room. During the day, the living room's leopard-stenciled rug, rock crystal tables, and an abundance of greenery are illuminated by light filtered through lavish drapery at the windows. The total effect of the light is enhanced by mirrors. Floor-to-ceiling shutters and heavily trimmed velvet drapery turn the master bedroom into a dusky haven. Light is transmitted through elaborately woven shades, either beaded or covered with several layers of gauze.

By opening up the first floor's attic in the Cohen's Bel Air home, Wilson extended the ceiling height by twelve feet in the 20-by-40-foot space. Uplights at the former ceiling line, as well as conventional table lamps, provide an abundance of reflected light off the soaring bleached white planes. The chandelier and candlelight provide the primary lighting in the dining room, carrying forth the warm glow established by the walls' rosy hue. "In the midst of a home's otherwise neutral palette, a richer hue in the dining room can be especially flattering to guests and can distinguish the dining experience from the rest of the evening," says Wilson. An exterior wall is shaped to echo the line of an arched window and exterior illumination reflects flattering ambient light into the room. In the bedroom, light reflected from uplights in the ceiling soffit mimics the warmth of daylight.

Explorations Resplendent

Wide-open Magic

Pebble Beach, California

Photography by Gil Edelstein and Mary E. Nichols

above Strip lights of the suspended bridge illuminate the beams, but not the glass, and mobile weighted-base Capri uplights throw patterns of light through the bamboo below. Projector fixtures dramatically frame art with light.

left Surface-mounted underwater fixtures in the front court-yard's reflection pool illuminate a bronze sculpture by Aristides Demetrio. Other landscape lighting is provided by low-voltage well lights and downlights.

Four ridge-long skylights and a bridge suspended through a glass atrium open the Schneider residence in Pebble Beach to the stunning Pacific environment, making most additional daytime illumination unnecessary. The challenge to lighting designer **Linda Ferry**, however, was how to illuminate the abundance of glass at night — when the windows seem like black voids and the skylights like black mirrors.

"To avoid reflection, every interior light source had to be individually placed. Illuminating the garden was of equal importance, so that the surrounding landscape could replace the ocean views which cannot be seen after nightfall," says Ferry. Working with architect **Charles Rose** and interior designer **John Schneider**, she used light to ensure that the strong exterior/interior concept did not diminish after sunset.

"A major goal was to not over-light, but simply replace the feeling of the natural light once night falls," says Ferry. The honey color of the Idaho quartzite floors, used generously to interpret the shoreline and visually anchor the residence on the site, assists the ambient light and reduces the light sources needed. So do the bleached wood ceilings, stark white walls, and even the "Amazon" gold granite in the kitchen, which keep the home bright long after the sun descends below the horizon.

above The soft hues of the quartzite floors and stark white walls assist the ambient light and reduce the number of light sources needed.

right Low-voltage downlights highlight the table, the ship in a glass box, and the interior of the upper cabinets. Uplighting is provided in the dining room soffit and from the Pfister halogen sconces by Boyd in the next room. The light sources are placed in pairs to enhance the architecture's symmetry.

THE PILGRIMS WAY
LOS ANGELES
ART AND CIVILIZATION

above Recessed downlights highlight the dining table and low-voltage spotlights accent the art. The table lamps by Machado contribute a warm, intimate feeling.

left Boyd Lighting's sconces illuminate the library's stylized coffer ceiling. A reading lamp allows the overall lighting to remain dim, making the most of the glow in the fireplace (one of six in the house).

above & opposite Recessed lighting fixtures in the ceiling and fluorescent strip lights in the soffit complement recessed incandescent downlights. In addition to illuminating kitchen work areas, the fixtures highlight the "Amazon" gold granite that recalls the appearance of coastal sand. Uplights create interest at the ridge-long skylight.

below In the exercise room, unobtrusive recessed fixtures provide sufficient light without causing glare on the video area and mirrors.

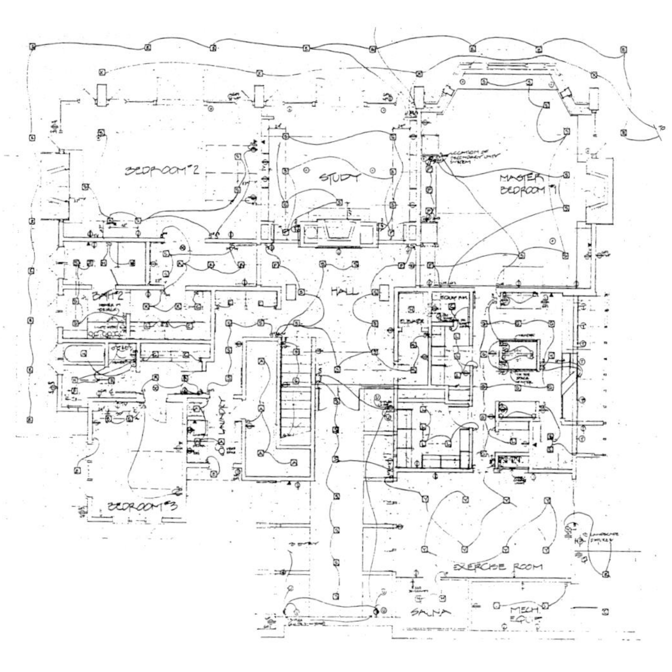

3 Homes, 3 Viewpoints

Rome, Italy

Photography by Edoardo D'Antona

top Stark-white walls punctuated by windows onto expansive views create a boundless atmosphere in the Colavita villa.
above A streamlined effect is achieved in the designer's own home.
left The Cimaglia residence is characterized by spotlighted artwork.

By day and night, the sky over the Colavita villa in Rome appears limitless, even when one is inside.

Designed by architect **Angelo Tartaglia**, the contemporary structure is characterized by large open spaces with expansive windows and one grand skylight with a view to the surrounding countryside. By night, however, pendant lighting makes the house itself look like a starlit sky.

Then look again. Two other Tartaglia-designed residences have completely different approaches to lighting. The Cimaglia apartment is illuminated primarily by spotlights that emphasize works of art. And in his own home, an apartment in one of the oldest sections of Rome, mostly experimental lighting (some of which had not yet been made available to the public at the time of installation) is different in every room. Light is used to center the eye on objects, not to be decorative in and of itself.

Says Tartaglia, "In each home, you need to be flexible, to use lighting differently for different things. Whether the mood is relaxing, dramatic or exciting in a contemporary way — like the architecture, the lighting has to reflect different points of view.

"That is the reason my lighting philosophy is not always the same. It depends on the place in which I am working."

In the Cimaglia residence, all lighting is accomplished through designer fixtures selected with as much care for their individual appearance as the objects they illuminate. Lustrous surfaces of stucco, painted Roman pink, lacquered black wood, and natural wood create drama and contrast throughout. At night, the extensive crystal collection reflects the designer lighting, enhancing the warm and intimate atmosphere. The main lighting fixtures are fabricated by: Fontana Arte (dining room); Martinelli (kitchen); Foscarini (hallway); Fontana Arte (living room); Venini (bedroom).

In the Colavita residence, the abundance of natural light flowing through skylights, windows and glass block renders the extensive lighting system unnecessary during the day. At night, the light sources, suspended at various lengths from the high ceilings, visually turn every room into a starlit sky. The effect is enhanced by the use of a reflective white paint throughout.

In his own residence, a small apartment in an old part of Rome, Angelo Tartaglia's goal was to provide a contrast to the location's antiquity by using newly introduced or experimental lighting fixtures set against highly reflective white-painted walls. All other lighting is halogen, used for spotlighting, uplighting and downlighting, and fluorescent lighting in the kitchen. Among the special fixtures included are ones manufactured by Martini, Fontana Arte, Flos and Reggiani.

HIGH ART

Dallas, Texas

Photography by ©1996 Ira Montgomery

To lighting designer **Barbara Bouyea**, this project was a mixed blessing.

The clients' first objective, highlighting their collection of contemporary and Asian art after it had been placed by interior designer **Cheri Etchelecu**, made it a choice assignment. Due to the high-rise location, minimal ceiling depths created a challenge for placing lighting, mechanical, and sprinkler systems. Architects **Bill Booziotis** and **Holly Hall** added highly configured wood ceilings, including various vaults. The ceilings were problematic for the precision lighting, which was required for the artwork. Some of the wood ceilings, in fact, had to be removable for access to other systems.

Nonetheless, Bouyea achieved her goal. Every work of art is illuminated through a combination of multipurpose adjustable accent lights, or surface-mounted fixtures where necessary, and linear shelf lighting. Sconces add ambient light. The particularly troublesome barrel vault in the dining room was carefully uplit by a small stainless steel architectural fixture, its shape echoed by the luminous ceiling Bouyea created in the kitchen.

All lighting is controlled by a Lutron four-scene preset dimming system, enabling it to provide illumination suitable for large events for one hundred people, a romantic interlude for two, and anything in between.

above In a hallway featuring a selection of the clients' impressive art collection, cold cathode cove lighting by Neotek illuminates the ceiling, highlighting the structural pecan element as another work of art. Reflected light draws the eye to the natural beauty of the limestone floor.

right The rooms glow with dramatic highlights provided by multipurpose adjustable accent lights, linear lighting on the glass shelves, and sconces which also add ambient light.

top In the dining area, the barrel vaulted ceiling is uplit by a simple, narrow stainless steel fixture by Baldinger. Accent lighting emphasizes the warm tone of the furnishings.

above In the master bathroom, small low-voltage downlights highlight the marble. Artemide wall sconces provide shadow-free illumination.

left Over the kitchen island, a luminous barrel-shaped ceiling which is backlit by numerous rows of neon, provides general lighting and keeps direct lighting off the high-gloss cabinets. Quartz UCL from Danalite provides task lighting.

opposite Accenting art and accessories with adjustable lamps helps this contemporary space avoid a cold, hard edge.

Geometric Adventure

Jeddah, Saudi Arabia

Photography by Samin N. Saddi

When **Zuhair H. Fayez** designed his own home, he hoped to establish an environment that would not only serve his family's needs, but also never lose its spell of excitement. Says the Jeddah-based architect with a flourishing practice throughout the kingdom, "As long as I live in it, I never want to get bored!"

To meet his requirements, Fayez assumed a task that required perfect planning. It had to satisfy his needs when he first designed it and had only one child. Presently it must accommodate his family of six children and eventually, it must adapt as they move out. At all times, the house needed to project a sense of being neither too big, nor too small.

"Perhaps just as important to me, being in the arts, is uniqueness and excitement," he says. "That means the design had to address the concept from solid geometric inspiration. It was imperative that it not be a novelty. And this translates into my desire to discover something new about the house as long as I live in it."

Fayez did indeed achieve his dream — realized through his fundamental concept of a series of octagons. Cut in half, shifted horizontally and stepped vertically, the shapes allow both natural and artificial light to reflect the forms' geometrical spirit.

Accenting the octagonal design, concealed uplighting at the recessed ceilings throughout the residence provides ambient light. A chandelier creates a dramatic focal point of illumination in the living room, while table lamps lend an intimate light to seating areas.

above In an intimate dining room, a mirrored wall and glass tabletop reflect light from above.

opposite (above left) In the formal dining room, the large crystal chandelier was designed to echo the shape of the long table. Etched glass panels transmit the hanging fixture's sparkle into the adjacent room.

opposite (above right) In the living room, a circular soffit radiates light from a crystal chandelier.

opposite (below) In the office and library, the cabinets are lit from within to accent objects and provide general ambient lighting. A decorative table lamp provides a residential feeling.

Flying High

United States of America

Photography by Mary Nichols

The creation of any environment for people in transit, where space is at a premium, presents many challenges to the interior designer. Perhaps none could be greater than outfitting a private jet, which was the task presented to New York's **Joseph Braswell**, who turned this Boeing 737 into a place of psychological as well as physical comfort.

"Foremost in an airplane, a climate of safety and tranquility is imperative," he says. "Equally important is avoiding a feeling of claustrophobia. It is also necessary to give the traveler a sense of well-being, expectancy and glamour."

To that end, generous provision was made for relaxation, dining, conferences and sleeping for extended travel time. Materials throughout — wool plush, Thai silk, glove leather, wool limousine cloth, wool carpeting, suede wallcovering and bird's-eye maple paneling — are in soft muted tones that create an aura of harmony.

Beyond color and form, Braswell considers lighting a powerful tool. To achieve a flexible lighting system for both day and evening hours, he used a mix of ambient and task lighting. Considering the restricted space and its use, the effect is astonishing.

above The aft lounge is dedicated to business meetings but also allows for relaxation. Low-voltage pinpoint spots accent tables and chairs. Perimeter strip lighting is repeated at the windows.

right At the aft bulkhead, Lucite mirror paneling is etched with radiating inscribed lines. Illuminated on the peripheral edges, they create a fiber-optic effect and expand the sense of space. Low-voltage pinpoint spots are placed for tabletops, chairs and sofa. Decorative Lucite lamps flank the sofabed to accentuate the fiber-optic theme.

In the central lounge, the overhead section is punctuated with circular skylights which are edge-lighted with fluorescent tubing. The surface is sheathed in Lucite mirror, producing an illusion of natural daylight. Low-voltage pinpoint spots accommodate the tables, chairs and sofabeds. Perimeter strip lighting illuminates the curtained windows. Reading lamps and wall-mounted fixtures create a residential feeling.

Light Dressing

Southeastern Brazil

Photography by Tuca Reinés

Simplicity with endless possibilities was the goal of São Paulo's **Arthur de Mattos Casas** in designing this home for a couple who decided to pare down. When their sons married and moved away, their former large residence seemed inappropriate to their current needs. Their preference now was for a smaller, more compact and practical space.

To scale down without a sense of confinement, de Mattos Casas created an abundant sense of flow between the various living areas, and selected and designed furnishings that are understated, highly edited forms which do not encumber the space.

"Yet, as important as the interior architecture and furniture were in achieving our goal," says de Mattos Casas, "the third element — lighting — carried equal weight."

Major concerns were balancing the intense external daylight with the interior artificial lighting, using materials that dramatically reflect light, as well as spotlighting furniture and art to provide as many points of interest as might be offered in a larger space. To give the clients the widest range of lighting options, de Mattos Casas made it possible for them to regulate the independent circuits from one command center.

Notes de Mattos Casas, "Particularly in scaled-down, simplified spaces, the illumination becomes fundamental in 'dressing' the rooms and providing variety."

above At the entrance hall, a marble pattern in the design of a welcoming runner has a high-gloss, glass-paste finish which is enhanced by the overhead use of tightly focused, energy-conservative (50 watt) dichroic spotlights.

right Natural wood shutters regulate the intense luminosity of the windows' northern orientation without losing visual contact with the external environment. The interaction between the daylight and the dramatic luminosity of the spotlights creates a wide range of moods.

above In the master bathroom, vellum luminaires lend a soft sidelight at the mirror. The 100-watt incandescent spotlights have a flattering effect.

above left Pull-down window covering in a natural tone softens the abundant daylight in the master bedroom. Dichroic spots provide task as well as accent lighting at the dresser, which is also used as a desk. The standing fixture, made of paper and chromed steel, is designed by Arthur de Mattos Casas, fabricated by Casas Design.

left Custom luminaires of translucent Carrara marble flank the mirror in the guest bathroom. A dichroic spotlight provides additional illumination and maximizes the silver basin's brilliance.

opposite In the dining room, concentrated-focus dichroic spots are directed primarily toward walls, furnishings and art. A paper lantern designed by Isamu Noguchi and manufactured by Akari Associates stands in the living area beyond.

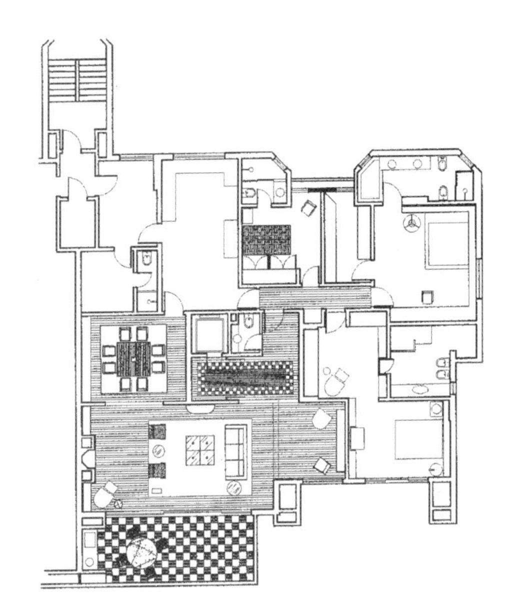

Total Transformation

Los Angeles, California

Photography by Derek Rath

Designed by a European architect in 1979, this residence was given a total transformation by lighting and interior designer **Bruce Liebert** of Avatar Design. Formerly a traditional architectural understatement, it now provides a clean, contemporary and highly dramatic setting for the owners' extensive entertaining.

Almost no areas escaped change. Rooms were moved, greatly enlarged or rebuilt. The large new kitchen has the best engineering available with new Bulthaup cabinetry fitted with halogen underlighting, providing brilliant, color-accurate task lighting on the granite counters. The living and dining rooms benefit from the removal of sliding metal and glass doors which were replaced by French wood doors opening to the newly re-landscaped and illuminated garden. In the living/dining area, existing fixtures were covered with custom monopoint canopies of brushed aluminum to hold Flos Lucy fixtures. Halogen cove lighting was placed on top of the structural beams to fill the huge vault with reflected light. The dazzling, 450-square-foot, granite-clad master bathroom has a Japanese-style soaking tub, a steam room that can accommodate eight people, and access to the garden.

Throughout, the completely redesigned lighting system uses the most current technology to lend further function and visual excitement.

The dining room's 20-foot peaked ceiling allowed the use of the sculptural Oh Mei chandelier by Ingo Maurer. Its silver-leafed paper panel, halogen lamp and steel squares are suspended on a slender metal frame by airplane cables.

The choice of a low-voltage exposed-cable lighting system in the remodeled kitchen was determined by the high pitched ceiling and lack of attic space to accommodate the wiring. This Translite Cablelite system uses the Byrdy and Luna fixtures for a combination of task and ambient lighting. Two pairs of Translite Vision monopoints, installed back-to-back on a beam, provide task lighting at the sink work areas and also illuminate the plants at the greenhouse window. In the master bathroom, ambient lighting is provided by ceiling-mounted, Fontana Arte Sillaba, low-voltage fixtures designed by Achille Castiglioni. A Leucos Selis 45 wall sconce is installed over the tub. The mirrors are lighted from behind with strips of halogen cove lighting to provide illumination in the sink areas. Reggiani Splash waterproof shower fixtures are located inside the stall shower and steam room.

Dazzling Simplicity

Home/Office Showcase

Dallas, Texas

Photography by Robt. Ames Cook

"It's homey . . . with a light show!" **Craig A. Roeder** says of the two-story 1904 house in Dallas that serves not only as his home but as headquarters for his international lighting design firm.

Located on the first floor, the office is used as a showcase and design studio. Cable lighting, multiple lamp types, shelf lighting and spotlights illuminate Roeder's exotic collection of art and artifacts. But perhaps his most stunning work of art is yet another form of lighting — two rows of cold cathode hidden in coves, on a preset, computerized dimming system. The cathode ranges in nanometers from 2700 degrees Kelvin to 3500 degrees Kelvin, enabling Roeder to demonstrate a broad range of lighting types to clients.

Specially designed wood moldings conceal the cold cathode- and neon-lighted coves. The ceilings of the second floor have double coves, created by Roeder and architect **Robert Oakes** during renovation. Their additional three-foot depth provides a spectacular visual effect. Illuminated with neon in the three primary colors and set on a computer controlled by a sixteen-button dimming panel, the ceiling colors can be changed for wide-ranging moods and effects. In fact, when the lighting hues are a certain purply blue, the ceilings look so ethereal that people mistake them for skylights.

Beyond Craig A. Roeder Associates' welcoming reception room, the carefully groomed garden is brilliantly illuminated so as to continue the ample views at night. Recessed accent lighting is by Edison Price. In the living room and all other major rooms of the second floor, Roeder created a double cove by pushing the ceiling up three feet. Dimmable, low-voltage downlights and recessed neon strips showcase different modes of office lighting in the architectural workroom.

above The bathroom sconces by Winona Lighting highlight the unusual taupe-colored marble.

right So as not to compete with his precisely pinspot-lighted collection of art and artifacts, Roeder painted all walls in the home and office the same gray-green. The coves are a lighter version of the same hue, and the ceilings are a light powder blue.

previous spread At night, colors cross-fade for a limitless range of moods in the upstairs dining room. During the day, natural light fills the room.

Summertime Magic

Karuizawa, Nagano, Japan

Photography by Hiroshi Shinozawa

above The design of the cubic, translucent, ivory-stained glass fixtures on the balcony was achieved after many meetings between the architect and lighting designers using models for location and form. Light Creation Inc. fabricated a small, intimate type of fixture desirable not only for large gatherings, but also for afternoon reading and relaxed conversations after a sauna.

left At the corner of the living room, twin pendants made of Japanese Kraft paper echo the dining room's hanging lights. All are designed by Akari Lisa Ishii. The wall-wash on the drawing, spotlights at stepside, and uplights can be dimmed to enable the glow from the fireplace to dominate at night.

Lighting designer **Motoko Ishii** wanted her family's summer cottage, located in a distinguished area of Japan's historical and popular mountain resort, Karuizawa, to be relaxing and cozy. A minimal palette of unadorned wood and white plaster lends a warm but simple summer-in-the-mountains atmosphere to the exterior, and the interior spaces pay homage to the nature outside as well.

Working in tandem with her daughter **Akari Lisa Ishii**, who designed the lighting, Ishii's goal was to emphasize the natural feeling of architect **Akira Ozawa's** architecture. Lighting fixtures and distribution — pendant lighting for focal warmth and small spotlights to accentuate steps, partition walls and art — favor architectural thought while continuing the setting's relaxed, comfortable atmosphere, from day into night. For example, footlights located below the windows in the entry passage, to which Motoko Ishii refers as "star-looking lights," are intended to illuminate without glare when looking up through the windows toward the stars.

In addition, to create a feeling of being in nature and far from urban life, most of the lamps can be controlled by dimmers, maximizing the enjoyment of the glow emanating from the fireplace, lighted candles and even the moon.

The dining room pendant light is augmented by downlights at the bay window which provide exterior light; spotlight wall washing on the inclined ceiling of the entrance hall; spotlights on focal points of art; floodlighting which pours through the skylights; and, for special occasions, candlelight.

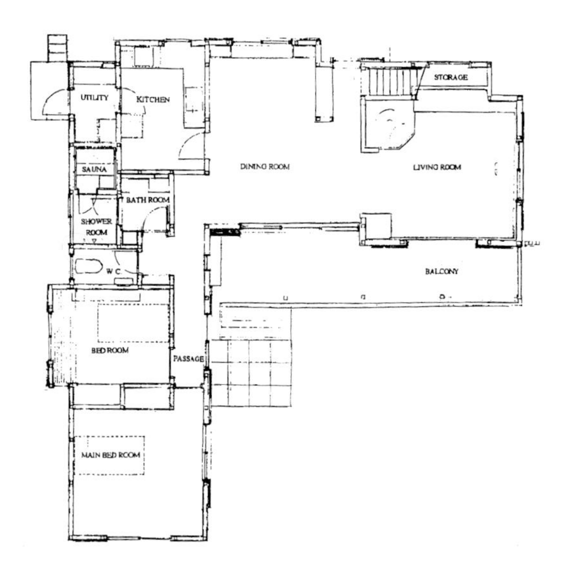

Bold Attitude

Los Angeles, California

Photography by Derek Rath

above A contemporary circular chandelier, made of Murano glass by Barovier & Toso, uplights the dining room's hand-troweled plaster ceiling. Room edges and wall art are defined by adjustable recessed low-voltage fixtures with wide-flood *MR 16* lamps.

left In the living room, adjustable recessed low-voltage fixtures are used for perimeter lighting, and wide-flood *MR 16s* highlight the fir ceiling and soft slate-blue plaster. These light sources define the edge of the space and take advantage of the reflected light without detracting from the view. A halogen beam shows off the rich wood coffee table base and objects on its glass top.

In order for the exciting colors and rich textures in this indoor/outdoor California home by architect **John Powell** and interior designer **Coryne Lovick** to be effective in the evening, color-corrective lighting became paramount.

Lighting designer **Bruce Liebert** of Avatar Design paid special attention to the home's varied shapes, tones and surfaces, selecting the best lighting for both individual areas and overall effect. Flooding surfaces evenly with light and using halogen lighting whenever possible accomplished the desired look — maintaining the forms' true colors and the impact of their mass.

In addition, the home's decidedly contemporary casual feeling is echoed by the lighting's unobtrusive yet ubiquitous quality. In the kitchen, for example, four light sources provide maximum utility: recessed fixtures for general lighting over the passageway; small-aperture recessed fixtures for task lighting; and an adjustable, miniature low-voltage system under the cabinets for sparkle at the granite counters. All of these help to achieve the concept of non-intrusion, with reflected light enveloping the occupant while minimizing shadows.

For easy-to-use control of this elaborate lighting system, Liebert divided the house into five zones and provided each with a Lutron Grafik Eye dimming system and an AMX linking system.

opposite above The entry is illuminated by skylights and by halogen cove lighting for a subtle, pure and even wash. Exterior lighting at the glass curtain is provided by recessed 45-watt halogen reflectors to accent the texture of the slate walkways.

opposite below In the kitchen, Liebert used a combination of recessed fixtures for task and general lighting, and low-voltage lighting under the cabinets.

left The breakfast room provides a warm gathering place at night too. An adjustable recessed fixture fitted with MR 16 lamps accents the fireplace.

below In the master bedroom, a freestanding wall dividing the bedroom from the closet provides an ideal location for halogen cove lighting that lends a reflected glow to the room. Above the bed are individually controlled, adjustable recessed low-voltage fixtures.

Classic Understatement

Mexico City, Mexico

Photography by Lourdes Legorreta

"In Mexico, we used to say that the most difficult work for an architect is his own house," says **Víctor Legorreta**. "Maybe it is because you have to deal with the most difficult client of all — your wife — or maybe because it is the one project where you are most free to explore all your ideas and use all the different materials and forms you have seen in your life."

The idea for Casa Víctor and Jacinta was a series of simple, clean, contemporary spaces that would be classically pure in form but not at all traditional. Instead of formally defined spaces, the dining table was to double as a worktable, the living room as a workplace, the reception area as a living room.

The play of light throughout the interior was a major consideration. All main rooms needed to face a central courtyard to take advantage of the sun, with windows further opening every space to the sky. Additional lighting, planned by lighting designer **Ing. Héctor Nieto**, is highly natural in feeling and, like the architecture, highly edited to further emphasize the sense of comfort and to turn the house itself into a play of shadow and light.

above Uplighting emphasizes the warm ochre color along the entry hall.

right The sunken living room/workplace accentuates sunlight. Directed ceiling lights accent the topmost shelves.

opposite right The small square shape of the peephole in the front door is the first of many such openings throughout the house, which add architectural interest through the play of light.

above The reception/living area looks toward the patio/courtyard through an amplitude of glass doors and windows.

opposite above Handmade tiles cover the kitchen ceiling. Illumination is provided by a central skylight.

opposite below In the bathroom, one intense uplight emphasizes the rich stroke of color and provides ample ambient illumination for the entire room.

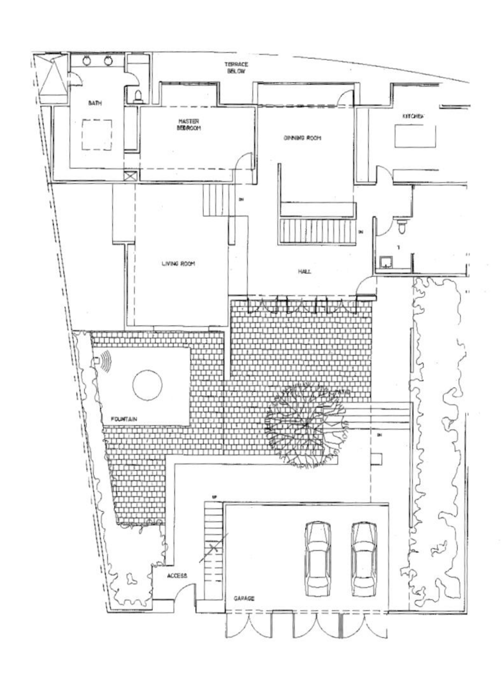
TERRACE
BELOW
BATH
MASTER
BEDROOM
DINING ROOM
LIVING ROOM
HALL
FOUNTAIN
ACCESS
GARAGE

Luminous Evocations

Santa Monica, California

Photography by Hugo Rojas

above In the living room, double-hung windows are handcrafted by Pedro Bolanos.

left The sunroom's cupola, with skylights, handcrafted by Pedro Bolanos, visually sculpts the architectural forms with natural light. Custom lamps serve interior tasks while exterior lighting delineates the architectural enclosure.

A blend of traditional and contemporary architecture, the home of **Geoffrey Scott** provides an ideal laboratory in which to illustrate his multidimensional talents. Scott not only designs architecture, interiors, furniture and graphics — in almost every case he also evokes a sense of the future even as he re-creates the past.

Working in tandem with his associates **Nikki Chen** and **Josue Vasquez**, he created a sequential flow from one comfortably scaled interior space to the next. Rooms are traditionally suited to function, yet relate to the adjacent spaces through Scott's commitment to balancing modern design influences with traditional details and elements. Every space features his sculptural light forms, with some serving as table and floor lamps for task lighting. These romantic vignettes include artful light forms that turn history on its head, yet still perform effectively.

Using quarter-rounded details and radius corners for the lath and plaster construction and painting every square inch a linen white, Scott enables daylight to flow continuously without sharp shadows and lines. In the evening, accent lights enhance function, and focal-point lighting lends even more drama to Scott's artistic expressions.

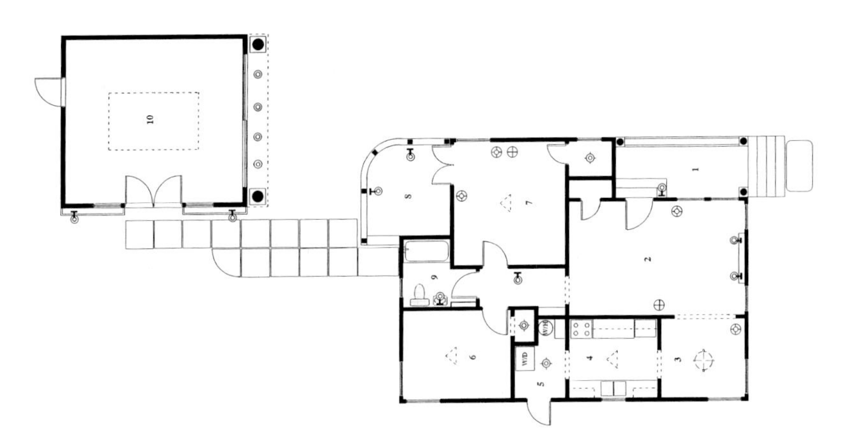

above In the master bedroom, the window and handcrafted French doors admit light that is complemented by the incandescent light of floor and table lamps designed by Geoffrey Scott and Associates. Porch lights create exterior interest.

opposite above Vesta wall sconces, an Angel table lamp and Globe floor lamp provide mood as well as task lighting in the living room.

opposite below A dimmable pendant light by Aerie turns the dining room's illumination into sculpture while establishing the table as an important focal point. The table lamp by Geoffrey Scott and Associates provides an accent.

Material Transformation

Washington, D.C.

Photography by ©Julia Heine

One of six speculative co-op houses, this multilevel home was built simply in terms of details — but it had good bones.

When his client asked him to elaborate on the drywall interior, architect **Mark McInturff** changed little in terms of volume or square footage, but much in terms of materials. He created an interior architecture of wood, aluminum, stone and fiber — an envelope contrasting warmth, coolness and translucency.

Triangular maple columns with aluminum reveals, internally lit and topped by tubes made of a fiberglass product that resembles Japanese rice paper, define the entry area and set off the living room. The columns continue up the stairs and mark the boundary of the new kitchen. Dining, kitchen and living areas are linked by a dropped ceiling plane which supports a sixteen-foot V-shaped light fixture made of aluminum and fiberglass. The axis (formed by the ceiling plane) ends in the living room, where the windows are framed in a new blue wall plane. The large windows can be covered in a variety of configurations by a system of sliding panels, steel framed and covered in the same translucent fiberglass material.

On the outer wall, two slender triangular bay windows project into the surrounding greenery, and a tall, narrow window accompanies the stair as it ascends four levels.

Light columns flank the tall triangular windows that echo the columns' shape and project beyond the wall plane into the trees beyond. Sliding screens at the large front windows, illuminated columns, and sandblasted glass at the front door provide both privacy and light. A curved wall carrying a pair of sandblasted doors separates the dining room from the kitchen. Walls are punctuated by illuminated columns.

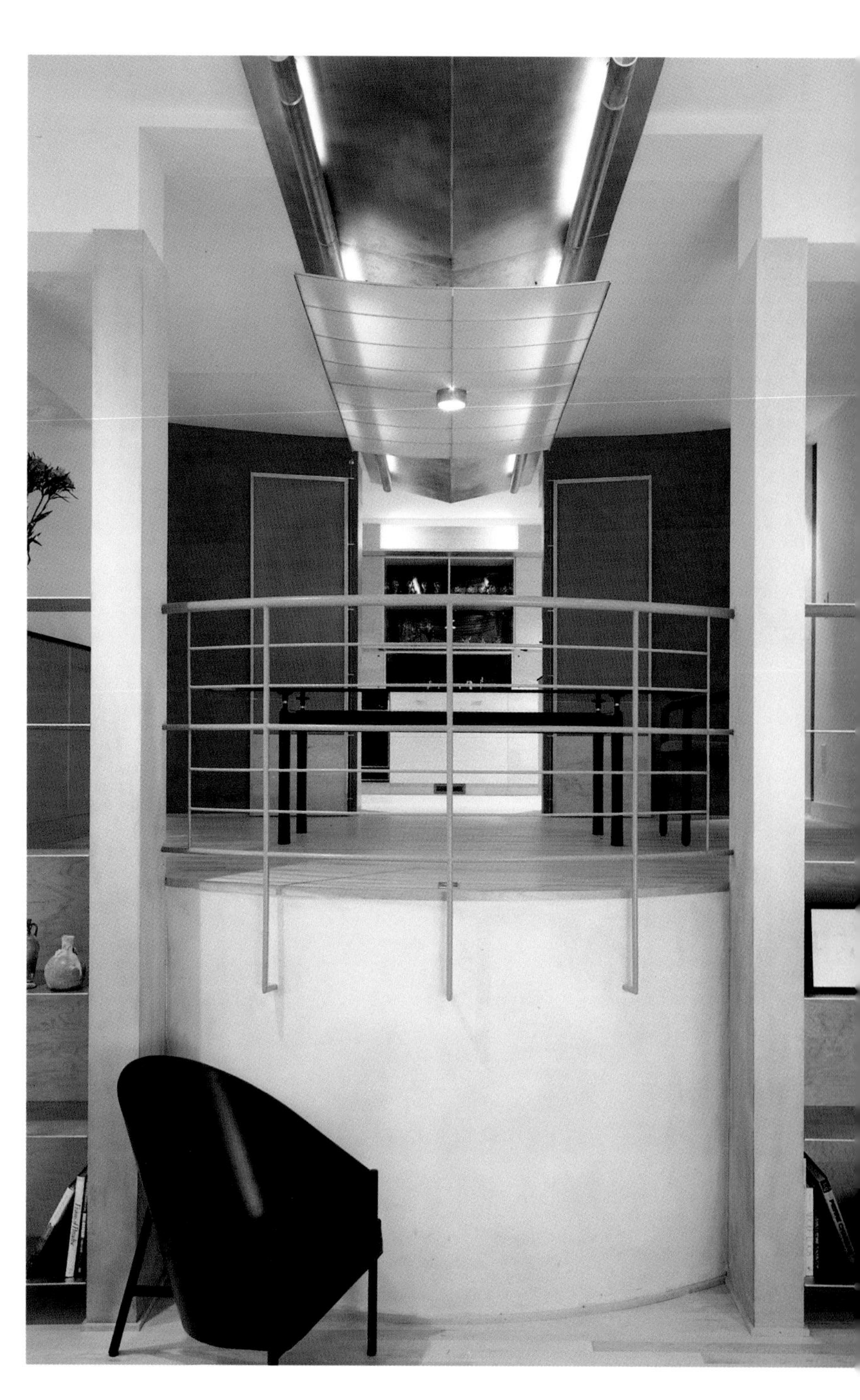

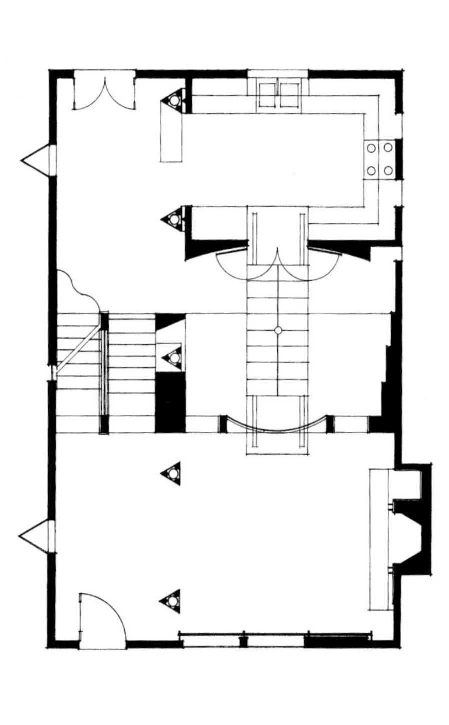

above In the kitchen, downlighting provides general illumination, while soffit lighting casts an ethereal glow across the ceiling. Light columns in the foreground are fabricated from translucent fiberglass that has been torn to provide an irregular edge, then rolled into a tube and riveted.

opposite above Illuminated columns define the circulation space, while the long ceiling fixture marks the axis running through living, dining and kitchen spaces.

opposite below The black granite and metal of the fireplace were chosen for their reflective qualities, as was the glass shelving for the display niche and aluminum lining.

More For Less

Sausalito, California

Photography by Barry Brukoff

Le Corbusier's architecture has long inspired the work of **Barry Brukoff**, as is evident in the way he altered the shape of his own condominium, replacing its nondescript boxy shape with simple but sweeping curvilinear and angled forms.

New plasterwork introduces gentle, sculptural arc forms. Slabs of slate give visual stature. Expansive planes of maple woodwork provide warmth and contrast. And throughout the residence, all spaces are punctuated by classic contemporary furniture and exotic acquisitions garnered by Brukoff on his intensive international sojourns.

Brukoff gutted and remodeled the 1,000-square-foot residence for $50,000. In keeping with the modest budget as well as his non-decorative design approach, the lighting is as simple as possible and — not surprisingly, as Brukoff is a lighting consultant as well as architect, interior designer and fine artist — equally effective.

The apartment is a vehicle for Brukoff's art collection, including his own painting, sculpture and photography. Every room is accented by a mix of cove, niche and track lighting in ways that highlight the objects and also visually expand the sense of space.

"A spacious feeling is never the result of additional square footage alone, and is always dependent on the quality of the architecture, interior design and light," says Brukoff. "To me the three are inseparable."

top In the dining area, a low-voltage minitrack lighting system with 20-watt narrow spot lamps is used to add sparkle and focus attention on the tabletop and painting.

above Track lighting illuminates the sculptural arc of the new plaster fireplace wall and provides light for the lounge chair and table. Uplighting in the soffit above the window extends the sense of spaciousness.

above right Low-voltage track lights accent artwork in the living room, while a 20-watt accent light located under the sculpted wooden horse provides dramatic uplight.

above A display niche designed to hold a Japanese raku sculpture of a kimono is illuminated by recessed puck lighting with frosted lens.

below left In the study, low-voltage Xenon striplighting hidden on the top of the bookcases provides indirect room lighting.

above In the bedroom, light from a wall sconce combines with natural illumination to produce an even glow.

right In the powder room, an Italian cast-glass sconce fitted with 150-watt halogen lamps produces glowing burgundy accents as well as enough uplight to illuminate the entire room.

opposite The rich hues of the natural materials palette — Kashmiri slate tile, Chinese slate and maple — benefit from SP 30 color temperature fluorescent lamps fitted into the valance which provide warmth.

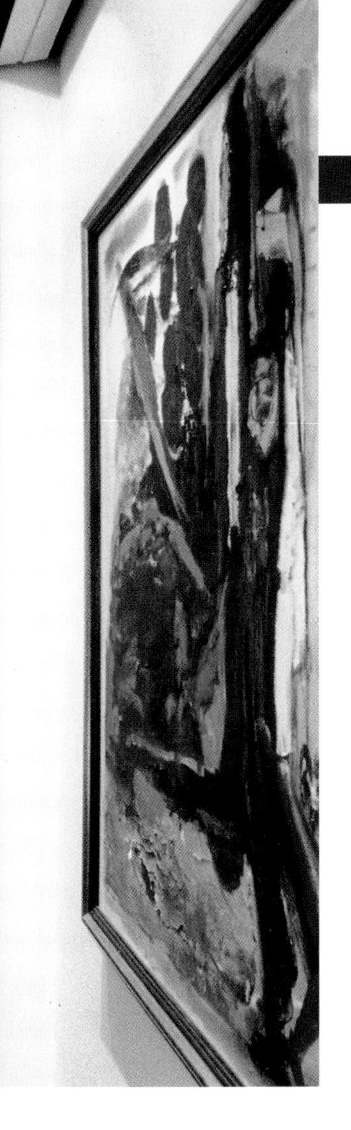

Intimate Harbor

Alsmeer, Holland

Photography by Frank Bonnet

Illumination at the exhibit hall addition is provided entirely by *MR 16* 20-watt ceiling-mounted accent lighting, placed at 24 jackpoints. Downlights, combined with accent light, illuminate artwork in the living room.

When their clients asked Architectburo Visser & Beerman's **Max Zimmerman** and lighting designer **Aleksandar Rublek** to renovate their home, their desire was to create a more spacious feeling — not only for living, dining and entertaining, but also for the display of art.

In addition to the installation of skylights, carefully placed task lighting, and decorative fixtures, the primary solution was to extend one side of the house for an exhibition area. It is enclosed by glass on the side overlooking the garden, and the exterior glass walls are flanked by display cases, to provide a sparkling presentation area for the homeowners' collection of glass art. The new interior walls of white-painted plaster provide additional space for paintings and prints.

Rublek used only accent lighting for the art itself — an arrangement of twenty-four individually operated jackpoints, each one controlling a halogen 20-watt lamp. When only ten are illuminated, enough ambient lighting is provided for general use. The benefit of having each one individually operated is that the presentation of art objects can be altered with ease.

"To me, since it is the only changeable factor," says Rublek, "light is architecture's fourth dimension."

above The kitchen is illuminated by four downlights in glass shrouds.

left Downlights in the dining area are augmented by accent lighting.

opposite above The bedroom is illuminated by the bathroom lighting, which is transmitted through the glass partitions. Reflected light from the wall opposite the bed, illuminated with recessed adjustable spotlights located in the ceiling, provides additional ambient light.

opposite left In the windowless entry, a collection of sweater storage boxes painted especially for the owners by Dutch artists is illuminated by newly added skylights during the day, and by *MR 16* 20-watt track lighting at night.

opposte right In the master bathroom, halogen side lighting combines with overhead incandescent lighting to provide full illumination at the mirror.

Kunstwerk

ALFRESCO — BY NIGHT!

Pebble Beach, California

Photography by Philip Harvey Photography, San Francisco and ©1994 Douglas A. Salin

The architect **Lee von Hasseln** designed many homes for herself, including second homes in France and Mexico. Her residence in northern California, enlivened by her personal finds throughout the world, reflects the international experience of those earlier projects. Yet the aspect of her favorite travels that she wished most to impart to this home comes down to this: sophisticated simplicity.

When von Hasseln asked **Linda Ferry** to professionally light her home — with as few lights as possible for a home that benefits from twenty-five skylights — Ferry had her work cut out for her.

With narrow limits placed on the lighting design, an approach was devised to create illumination that conveys a daylight quality. Soft, incandescent lighting which simulates ambient daylight was used to illuminate the architectural elements. A minimal number of sources was used, keeping the feeling simple so as not to draw attention to the technology. Even the light sources hidden in the deep wells of key skylights re-create the effect of the skylights in the day without being obtrusive. The result is design imbued with the feeling of completely natural luminescence — even after nightfall.

above Reflective recessed lighting by Capri which illuminates the doorway, and spotlighting on sculptures in the window niches provide an understated welcome to the architect's home.

right Highly reflective halogen fixtures with an ability to throw a wide beam emulate the skylight's aspect during the day, filling the room with a soft and pleasant level of brightness.

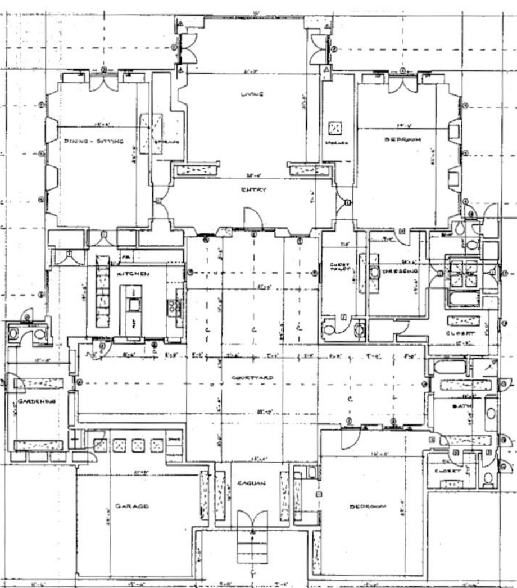

above In the sitting room, focal lighting accents art, creating a rhythm of contrast with landscape uplighting seen through the window.

right In the garden room, simple downlights gently wash shelves and provide efficient and cost-effective illumination for tasks.

Lustrous
Elegance

Whimsically European

Beverly Hills, California

Photography by Derek Rath

When Cippananda Interior Design's founder **Gabriella Toro** moved from Italy to Los Angeles, she brought along her sense of good materials, finishes and lighting. In this residence, which she completed with lighting designer **Lori Bush**, that feeling shines.

The Mediterranean-style house is owned by a single man who entertains often, both formally and spontaneously. The house needed to be comfortable and also provide a sense of high drama — especially at night.

The spacious living room is lit by downlights to ensconce guests in a warm glow of reflected illumination, yet not interfere with the view. The dynamic dining room, hugged by Pompeiian red-plastered walls, is illuminated by a star-motif chandelier. The media room achieves an even greater balance of function and intimacy through a combination of light and texture.

"I promised my client, who owns three cats, that I would design him a 'cat room,' where the sense of touch would be as important as the sense of sight," says Toro.

The tactile qualities of delicate silks and heavy tribal pieces are illuminated by candles, Giacometti and Philippe Starck lamps, and a pair of custom-made wall sconces, tall enough to demarcate a sense of passage into the room — its burnished, Venetian plaster walls further conveying the "touch" that Toro guaranteed.

In the media room, centrally placed incandescent downlights provide general lighting. Low-voltage accent lighting is on a separate circuit, enabling the general lighting to be turned off when the television is in use to reduce glare. The bronze table light in the corner has a rotating mirror, turning it into a flexible light sculpture.

In the dining room, an incandescent decorative lighting fixture with star-shaped cutouts is joined by low-voltage lights that accent art and downlighting to provide a sense of intimacy while enhancing the overall design.

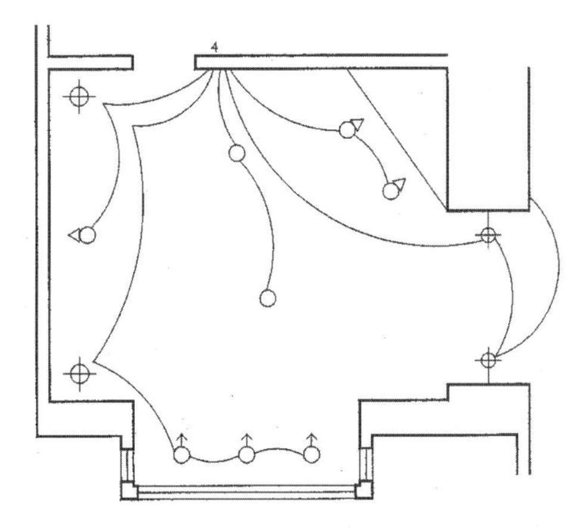

Rich Understatement

New York, New York

Photography by Mark Samu

Intense lighting highlights the sharp-edged architecture of this contemporary apartment in New York City designed by **Mark D. Stumer**, principal of Mojo Stumer Associates.

A duplex, this home is situated in a high-rise residential tower and consists of a living/working environment for a young, professional couple. The architect split the public/private areas by placing the main living area and studio at the upper level entry and the study/bedroom and master bedroom on the lower level.

"The couple wanted a clean, uncluttered, clearly organized, minimally detailed apartment that takes advantage of the breathtaking views. This drove the underlying concept behind the articulation of the entire project," says Stumer.

The simplicity of the upper-level plan allows for uninterrupted views throughout the apartment and increases the sense of spaciousness. The materials palette, also highly edited in terms of number and color, creates a frame for city views and allows them to remain paramount. In addition, the materials selected — hard and soft, cool and warm — create an interplay between formal and informal elements and serve as a non-competitive backdrop for art and accessories. Throughout, low-voltage, mostly *MR 16*, incandescent lighting is used as a strong architectural element, bringing to life the spirit of the forms but also having its own identity.

above A niche in the cabinetry is accented by low-voltage lighting directly over the flower arrangement.

right The architects enclosed the dining area, formerly an open terrace, with glass outfitted with roll-down window sunscreens that can be hidden in soffits above. The screens admit light but keep out ultraviolet rays. A custom fixture punctuates the ceiling grid, from which warm, white fluorescent light flows to art and walls below.

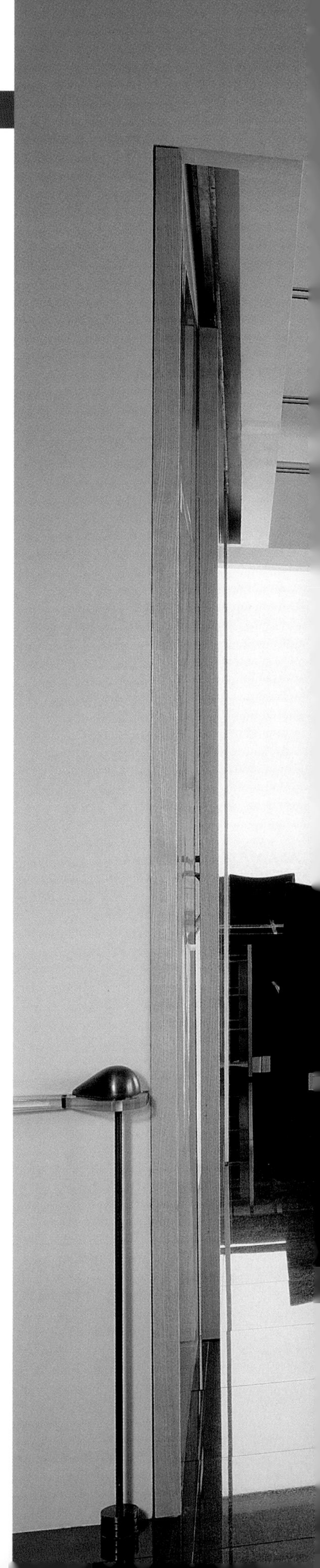

above In the dining area, a minimalist fixture of glass and steel helps maintain the extraordinary view.

left Downlights located in soffits and ceiling points highlight the richness and reflective quality of the highly edited materials palette.

previous spread The custom ceiling grid, uplighting the ceiling with warm white fluorescent light, does not compete with the cityscape.

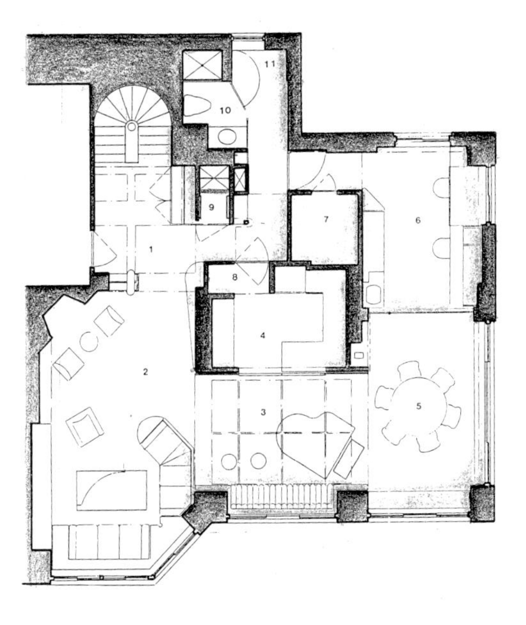

above The architects used downlights to duplicate the apartment's high level of daytime light at night.

right By day in the master bedroom, natural light provides the illumination, save for one decorative ceiling fixture. At night, downlights are used to bring out the shapes and shadows of the cabinetry.

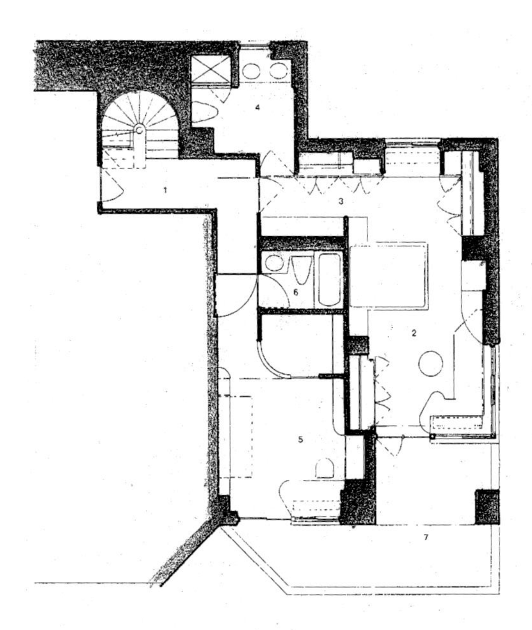

Guiding Light

Northern California

Photography by Mary E. Nichols and Douglas Salin

In the remodel of this desert home, light is the guiding element — literally. Wrapped around a central pool and courtyard, the layout called for some visual way to direct visitors from the entry to the main public areas opposite. The ideal beacon proved to be a custom hand-blown illuminated glass fixture by **Pam Morris** of Exciting Lighting.

Light serves many functions in these recreated spaces by interior designer **Helen C. Reuter** and Light Source lighting designers **Randall Whitehead** and **Catherine Ng**. Each function is integrated artfully — the home's existing architectural features hide indirect fixtures that provide ambient illumination and Morris's sculptures and other decorative fixtures shine in the foreground.

The home abounds in coves and other details that are tailor-made for indirect lighting. While adjustable recessed low-voltage fixtures highlight art and dramatize key architectural elements, ambient fill light creates a sense of comfort by adding a warm glow and softening shadows.

"We're big advocates of ambient lighting in homes," says Whitehead. "We always remember that in any interior there are three things we're lighting: the art, the architecture and the people. We start with the people."

above At the entry, Pam Morris's illuminated standing figure of nickel-plated iron holds a glowing pod of hand-blown glass. Additional lighting — adjustable recessed low-voltage lighting and low-voltage strip cove lighting — is controlled, by a Lutron Grafik Eye four-scene preset dimming system. Beyond is a large illuminated Georgia O'Keeffe painting, part of the owner's extensive art collection.

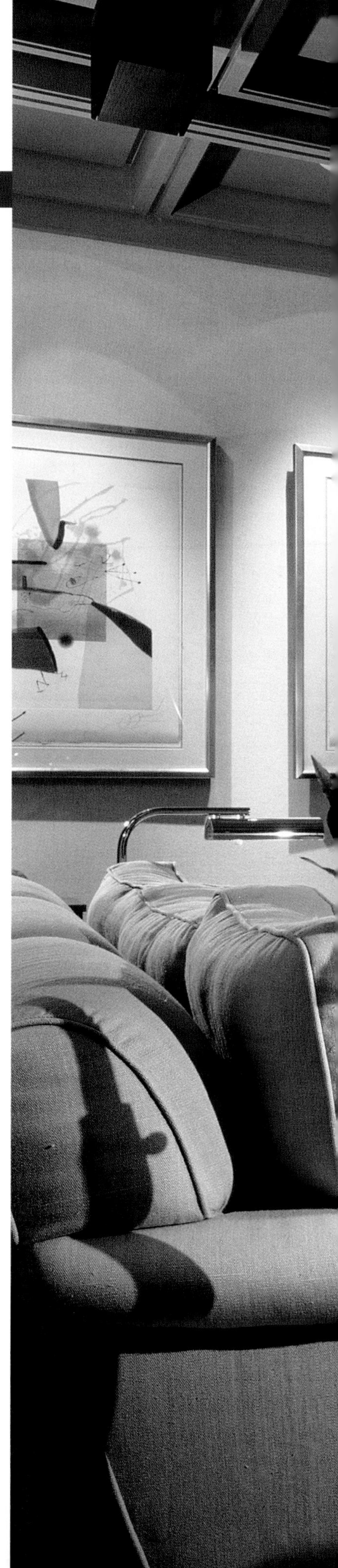

right A multifunction lighting system with a four-scene preset dimmer provides a variety of light levels in the media room. The system can be activated from a hand-held remote control, conveniently changing the room's atmosphere to make it suitable for watching movies, reading or playing games.

THE ABYSS

above True layered lighting — including spotlighting, indirect ambient light and decorative lighting — comes into play in the living room. Sconces and torchieres by Pam Morris were inspired by the sprouting forms of the agave plant.

left The dining room's commanding pendant light by Pam Morris is illuminated by a recessed halogen downlight. The floating paper-like art piece echoes the form of the table base below. Low-voltage strip lighting, placed below the cabinets and in the cove, punctuate the ceiling and the floor with light.

opposite The large master bedroom benefits from abundant ambient light, while its expansive area is subtly enhanced by additional lighting. Adjustable recessed fixtures add depth and dimension, and indirect lighting is installed above the soffit details. Four ceramic wall sconces by Boyd Lighting are mounted on the two pairs of columns, where no cove or valance could conceal a light source. The standard sconce uses an asymmetric reflector to distribute light for the specific use.

Comfortable Luxury

Lake Forest, Illinois

Photography by Bruce Van Inwegen

"Comfortable luxury" might seem like a contradiction in terms — especially when describing a contemporary Palladian-like villa ensconcing a grand contemporary art collection. But the clients wanted comfort, not a museum. And that fit the design philosophy of architect **Jim Olson**, principal of Olson/ Sundberg; lighting designer **Richard Renfro**, of Fisher Marantz Renfro Stone; and interior designer **Liz Fisher**.

"Although Richard Renfro and I met while working on the Seattle Art Museum, we believe that residential design should be more relaxed than a public building," says Olson, whose design for the home is as welcoming and open as it is monumental.

Renfro agrees: "In a museum, the concentration is primarily to illuminate the works of art. But in a home, the general ambience surrounding people is more important. So while we made sure we had proper illumination in the primary art locations, we worked with, instead of competed against, the natural light. In addition, the goal of the artificial light was illumination that would not only highlight the art but also be flexible and softly light people."

For example, adjustable fixtures enable art to be moved or added easily. And, instead of harsh downlighting and focused spotlighting, walls are washed with light, which reflects back into the room so that art is unpretentiously illuminated and people can be seen in a gentle, becoming light. In this home, "comfortable luxury" is no oxymoron.

above Despite its monumentality, the entry merges building and landscape in a way similar to Italian country villas and the Prairie traditions of Frank Lloyd Wright. Sweeping terraces and powerful horizontal planes place the house squarely in the Midwestern tradition of space.

opposite In the living room, natural light and downlights create a clean glow for viewing art.

Comfortable Luxury

Lake Forest, Illinois

Photography by Bruce Van Inwegen

"Comfortable luxury" might seem like a contradiction in terms — especially when describing a contemporary Palladian-like villa ensconcing a grand contemporary art collection. But the clients wanted comfort, not a museum. And that fit the design philosophy of architect **Jim Olson**, principal of Olson/ Sundberg; lighting designer **Richard Renfro**, of Fisher Marantz Renfro Stone; and interior designer **Liz Fisher**.

"Although Richard Renfro and I met while working on the Seattle Art Museum, we believe that residential design should be more relaxed than a public building," says Olson, whose design for the home is as welcoming and open as it is monumental.

Renfro agrees: "In a museum, the concentration is primarily to illuminate the works of art. But in a home, the general ambience surrounding people is more important. So while we made sure we had proper illumination in the primary art locations, we worked with, instead of competed against, the natural light. In addition, the goal of the artificial light was illumination that would not only highlight the art but also be flexible and softly light people."

For example, adjustable fixtures enable art to be moved or added easily. And, instead of harsh downlighting and focused spotlighting, walls are washed with light, which reflects back into the room so that art is unpretentiously illuminated and people can be seen in a gentle, becoming light. In this home, "comfortable luxury" is no oxymoron.

above Despite its monumentality, the entry merges building and landscape in a way similar to Italian country villas and the Prairie traditions of Frank Lloyd Wright. Sweeping terraces and powerful horizontal planes place the house squarely in the Midwestern tradition of space.

opposite In the living room, natural light and downlights create a clean glow for viewing art.

above & left An abundance of natural light flowing in through the exquisitely detailed pavilion is augmented by a light-reflective neutral color palette and by downlighting that follows the curvature of the window space. It is dimmable to ensure that the landscape lighting is visible at night. The bronze figure within the pavilion is by Robert Graham. The bronze on the terrace is by Fernando Botero.

opposite In the dining room, indirect fluorescent uplighting in the cove ceiling creates a cloud-like effect. Small downlights throw light cascading down a copper-clad waterfall sculpture by California artist Eric Orr.

ROUND-THE-CLOCK SUN

Northern California

Photography by Alan Weintraub

The pyramid of the 24-foot-square living/dining room ceiling did not allow for recessed lighting, so cable lighting adds function and sparkle. Uplighting located behind a planter creates a chiaroscuro effect as it plays against the architecture's neutral white walls and pine ceiling. In the dining room portion of the living/dining area, *MR 16* halogen track lights highlight the tabletop.

"As pet-proof as possible!" was the owners' primary request. They did not want to be distracted by a lot of maintenance — notwithstanding their two large dogs.

Since both the husband and wife are in the computer world and work at all hours from home, another important issue was to sensitively control the flow of natural light into the rooms and to replicate it at night.

So when they asked the home's original architect, **Peter Duxbury**, to enlarge the house and he in turn suggested that **Barry Brukoff** design the interiors, hardy slate floors and waxed-stucco walls were chosen to provide the basis for a scratch-resistant materials palette. The materials also serve as foils to the architecture's overriding themes: a steeply rising, sandblasted-pine pyramidal ceiling, abundant natural light that flows through skylights and windows, and enough carefully positioned artificial lighting to augment dark spots by day and continue the sun's effect around the clock.

Brukoff, a painter and photographer as well as designer, used his artist's eye to echo the plan's strong simplicity by using just a few strokes of earthy colors for embellishment. To keep them looking drenched with sun throughout the night, he worked in collaboration with **Epifanio Juarez** on the lighting: a mixture of track, recessed, cable and surface-mounted halogen uplights.

above & below Windows and a wood-framed/sandblasted-glass front door were added to the formerly dark entry stairway, washing the waxed-stucco plaster staircase with natural light. A photograph Barry Brukoff took of a similarly hued wall in Morocco and published in his book *Morocco* serves as a form of trompe l'oeil.

opposite At the top of the entry stairs, track lighting flanks the perimeter of the skylight. Low-voltage reflector downlights are recessed into the beams of the vaulted ceiling and, on either side, the former tracklights were replaced with recessed *MR 16* low-voltage spotlights.

MEXICO

opposite In the newly constructed media room, cable lighting joins with natural light to reveal the warmth of European beech cabinetry, designed throughout the house by Barry Brukoff and Peter Duxbury.

left In the home office, two Italian surface-mounted halogen fixtures avoid computer-screen glare by reflecting light off the ceiling. A light shaft is fitted with low-voltage, *MR 16* track lighting (at top, not visible), making Italian stucco look like liquid sunlight twenty-four hours a day.

below In the master bedroom, lighting was introduced in the niches to highlight the glass vases. Above, *MR 16* spotlights illuminate the artwork. Atriculated reading lamps are located on each end of the headboard.

Glamour! Comfort! Culture!

Southern Mexico

Photography by Victor Benitez and Eitan Feinholz

Even when the structure is contemporary, homes designed by architects **Carlos** and **Gerard Pascal** are usually based on principles of classical architecture.

In these three residences in Mexico, the designers show how technology can be used to create a deeper awareness of one's environment and make one wish to stop and feel the coolness of the marble, caress the smoothly turned woods and gaze at the masses of stained glass which they have so elegantly illuminated.

"Classic order, balance and symmetry help to create the sequence and image associated with the glamour, comfort and cultural attitude for which we strive," says Gerard. "We believe the classical orders were a direct answer to people's need to have their surroundings reflect their culture, geography and history — aspects of living that enrich the human condition. Today, in this era of globalization, cultural identities are being lost and technology is taking over, eliminating to a great extent comfort, individuality and a connection to our heritage.

"In trying to eradicate this trend, we use light to elevate each room's atmosphere, and to enhance the quality and character of materials by defining shape, texture and color. Light also emphasizes the many focal points of art that we use in all our work."

In this first residence, the lighting design revives a cultured, refined lifestyle. A vaulted ceiling with stained-glass window, for example, evokes the work of Frank Lloyd Wright's Dana House, the reflection of light through the glass creating endless variations and original patterns on the wall. Floor lamps are from Koch + Lowy. A central skylight casts sunlight on a foyer's stone floor in a three-dimensional design. Completing the symmetrical composition are eighteenth-century Chinese cloisonné vases enhanced with reflected light, yellow stained-glass windows and chrome-and-glass French Art Deco-style wall sconces.

The classical flair of New York's hotels is recreated in grand style in this second home. Deep green wallpaper and green Tikal marble on the floor were used to darken an elevator foyer and increase the drama of the lobby beyond, which is illuminated by a huge skylight. Recessed lighting defines the arch. A small table lamp accents the wall's textured finish. By day, a skylight over the vaulted ceiling in the whirlpool bath admits natural light. A halogen spotlight accentuates the drama of a Buddha statue.

In this third residence by architects Carlos and Gerard Pascal, dimming systems allow for several moods in each room. The warmth of the spa room, with its rich color and trompe l'oeil painting, is enhanced by wrought-iron and glass sconces, and chandeliers from Murano, Italy. An Italian light pendant by Flos lends a contemporary contrast to a breakfast room's otherwise traditional, mahogany-paneled environment. Rich silks shine under halogen lighting, with different shades of red providing a strong character to the living room where the table lamps, uplights and downlights emphasize the forms of the artwork.

The classical flair of New York's hotels is re-created in grand style in this second home. Deep green wallpaper and green Tikal marble on the floor were used to darken an elevator foyer and increase the drama of the lobby beyond, which is illuminated by a huge sky-light. Recessed lighting defines the arch. A small table lamp accents the wall's textured finish. By day, a skylight over the vaulted ceiling in the whirlpool bath admits natural light. A halogen spotlight accentuates the drama of a Buddha statue.

In this third residence by architects Carlos and Gerard Pascal, dimming systems allow for several moods in each room. The warmth of the spa room, with its rich color and trompe l'oeil painting, is enhanced by wrought-iron and glass sconces, and chandeliers from Murano, Italy. An Italian light pendant by Flos lends a contemporary contrast to a breakfast room's otherwise traditional, mahogany-paneled environment. Rich silks shine under halogen lighting, with different shades of red providing a strong character to the living room where the table lamps, uplights and downlights emphasize the forms of the artwork.

Penthouse Magic

New York, New York

Photography by Philip H. Ennis

above General downlighting illuminates the stairwell leading up to the living/dining/kitchen areas on the apartment's second floor. The print at mid-landing is highlighted with a low-voltage system.

left The Halo decorative hanging fixtures help define the window and provide light at the love seat. An important aspect of these fixtures is that their light can be directed, ensuring that it will not create disturbing hotspots in the glass.

The primary goal in lighting this high-rise duplex apartment overlooking Manhattan was to preserve the view.

"We didn't want to overly illuminate, especially in the evening, in order to make the most of the city lights," says Mojo Stumer Associates principal **Mark D. Stumer**, who gutted the entire space to re-create the home's architecture and interiors as well as the lighting. "Key to our plan was minimizing direct perimeter downlighting in many areas. Most of the lighting comes from the cabinetry, resulting in a predominance of softly ambient, indirect light.

"Besides, people generally tend to overly light their homes. Highlights are important, and to achieve them, some areas have to be kept dark. No room should ever be uniformly illuminated."

To arrive at the right lighting balance in this and all his projects, Stumer first designs the plan and elevation, architectural detailing, and furniture placement. This enables him to work with the light as an artist would — using the interior as his canvas to create specific highlights, shadows and subtle variations. So integral is the furniture to his architectural and lighting plans that the end result is not flexible, nor is it intended to be.

Notes Stumer, "Move the couch and it would throw off the entire scheme!"

above One of the most overlooked surfaces for lighting placement is the stove's hood. Mark Stumer used it here to illuminate the stovetop as well as provide night-light for late snacks. The decorative hanging fixture over the breakfast table is by George Kovacs Lighting.

opposite In the dining room, the architect created a notched ceiling to contain the recessed light and echo the line of the dining table. Highlighting the custom cabinet, a custom-designed, bronze-sheathed vertical incandescent fixture serves as a sculptural element and may be used alone to illuminate the entire room.

above To emphasize the warm glow from the fireplace, only minimal low-voltage, incandescent downlights have been placed in the ceiling. Much of the room's lighting is provided by the low-voltage fixtures in the custom cabinetry. Vertical window treatments on the left and right of the fireplace block the view from a neighboring building to enhance privacy.

left At the second-story stair landing, a divider wall of anigré wood blocks the view to maintain an element of surprise upon entering the room. A soffit at the top of the wood panel contains low-voltage incandescent light, illuminating the ceiling and providing reflected light on the oak stripped flooring.

opposite above The master bath's entry is generously illuminated, sensuously back-lighting the tub that opens to the bedroom. Decorative hanging fixtures highlight the basins. Etched glass at the window preserves privacy while transmitting light.

opposite below In the master bedroom, the center of the room is kept relatively dark except for bedside task lighting. Directed downlights accent the art.

DIRECTORY

Architects and Designers

Architectural Illumination
Linda Ferry, IES
P.O. Box 2690
Monterey, California 93942
United States
Tel: (408) 649-3711
Fax: (408) 375-5897

Avatar Design, Ltd.
Bruce Liebert
3628 Holboro Drive
Los Angeles, California 90027
United States
Tel: (213) 663-5700
Fax: (213) 663-5800

Booziotis & Company
Bill Booziotis, AIA
Holly Hall, AIA
2400 A Empire Central Drive
Dallas, Texas 75235
United States
Tel: (214) 350-5051
Fax: (214) 250-5849

Bouyea & Associates
Barbara Bouyea, IALD, IESNA
3811 Turtle Creek Boulevard
Dallas, Texas 75219
United States
Tel: (214) 520-6580
Fax: (214) 520-6581

Brukoff Design Associates, Inc.
Barry Brukoff
480 Gate Five Road
Sausalito, California 94965
United States
Tel: (415) 332-6350
Fax: (415) 332-5968

Cheri Etchelecu Interior Design
Cheri Etchelecu, ASID
9400 North Central Expressway
Dallas, Texas 75231
United States
Tel: (214) 369-7489
Fax: (214) 369-3691

Cippananda Interior Design
Gabriella Toro
3795 Wade Street
Los Angeles, California 90066
United States
Tel/Fax: (310) 398-0207

Craig A. Roeder Associates, Inc.
Craig A. Roeder, IES, IALD
Robert Oakes
3829 North Hall Street
Dallas, Texas 75219
United States
Tel: (214) 528-2300
Fax: (214) 521-2300

David Martin & Associates
Wm. David Martin, AIA
P.O. Box 2053
Monterey, California 93940
United States
Tel/Fax: (408) 373-7101

Derek Porter Studio
Derek Porter, IES
6133 Kenwood Avenue
Kansas City, Missouri 64110
United States
Tel: (816) 444-0835
Fax: (816) 444-2198

Details, Inc.
Helen C. Reuter
1250 Jones Street
San Francisco, California 94108
United States
Tel: (415) 921-3236
Fax: (415) 921-0139

Diseños Integrales de Ingenieria
Ing. Héctor Nieto
Av. Revolución 1110
San José Insurgentes 03900
Mexico
Tel: (52) 5-651-2032
Fax: (52) 5-680-1599

Duxbury Architects
Peter Duxbury, AIA
382A First Street
Los Altos, California 94022
United States
Tel: (415) 917-3840
Fax: (415) 917-3848

Fisher Marantz Renfro Stone
Richard Renfro
126 Fifth Avenue
New York, New York 10011
United States
Tel: (212) 691-3020
Fax: (212) 633-1644

Geoffrey Scott & Associates
Geoffrey Scott
Nikki Chen
Josue Vasquez
2917 1/2 Main Street
Santa Monica, California 90405
United States
Tel: (310) 396-5416
Fax: (310) 399-5246

Hague & Associates
Dennis Hague
8687 Melrose Avenue
West Hollywood, California 90069
United States
Tel: (310) 289-1301
Fax: (310) 289-1302

John Powell Architects
John Powell
9028 Crescent Drive
Los Angeles, California 90046
United States
Tel: (213) 848-7130
Fax: (213) 848-7360

John Schneider Design
John Schneider
P.O. Box 1457
Pebble Beach, California 93953
United States
Tel: (408) 649-8221
Fax: (408) 372-9356

Joseph Braswell & Associates
Joseph Braswell, ASID
425 East 58th Street
New York, New York 10022
United States
Tel: (212) 688-1075
Fax: (212) 752-7167

Juarez Design
Epifanio Juarez
1181 Greenwood Avenue
Palo Alto, California 94301
United States
Tel: (415) 322-6550
Fax: (415) 322-7602

Julia Rezek Lighting Design
Julia Rezek, IALD
1345 Tolman Creek Road
Ashland, Oregon 97520
United States
Tel: (541) 488-4254
Fax: (541) 488-0779

Legorreta Arquitectos
Víctor Legorreta
Palacio de Versalles 285-A
Loma Reforma 11020
Mexico
Tel: (52) 5-251-9698
Fax: (52) 5-596-6162

Light Source Design Group
Catherine Ng, IES
Randall Whitehead, IALD
1246 18th Street
San Francisco, California 94107
United States
Tel: (415) 626-1210
Fax: (415) 626-1821

Liz Fisher Interiors
Liz Fisher
1320 Southwest 15th Street
Boca Raton, Florida 33486
United States
Tel: (561) 394-4292
Fax: (561) 750-1089

Lori Bush Lighting Design
Lori Bush, IALD, IESNA
621 South Broadway
Los Angeles, California 90014
United States
Tel: (213) 413-5831
Fax: (213) 622-1845

Lovick Design
Coryne Lovick
11339 Burnham Street
Los Angeles, California 90049
United States
Tel: (310) 475-5781
Fax: (310) 440-2926

McInturff Architects
Mark McInturff, AIA
4220 Leeward Place
Bethesda, Maryland 20816
United States
Tel: (301) 229-3705
Fax: (301) 229-6380

Michael Wolk Design Associates
Michael Wolk
2318 Northeast Second Court
Miami, Florida 33137-4506
United States
Tel: (305) 576-2898
Fax: (305) 576-2899

Mil Bodron Design
Mil Bodron
2801 West Lemmon Avenue
Dallas, Texas 75204
United States
Tel: (214) 871-7588
Fax: (214) 871-7587

Mojo Stumer Associates
Mark D. Stumer, AIA
55 Bryant Avenue
Roslyn, New York 11576
United States
Tel: (516) 625-3344
Fax: (516) 625-3418

Motoko Ishii Lighting Design, Inc.
Motoko Ishii, IES, IALD
Akari Lisa Ishii
5-4-11 Sendagaya
Shibuya-ku, Tokyo
Japan
Tel: (81) 3-3353-5311
Fax: (81) 3-3353-5120

Olson Sundberg Architects
Jim Olson
108 First Avenue South
Seattle, Washington 98104
United States
Tel: (206) 624-5670
Fax: (206) 624-3730

Ozawa & Architects
Akira Ozawa
1-18-4 #301 Shinjuku
Shinjuku-ku, Tokyo
Japan
Tel: (81) 3-3226-9303
Fax: (81) 3-3226-9360

Pascal Arquitectos
Carlos Pascal, IIDA
Gerard Pascal, IIDA
Atlaltunco 99
Tecamachalco, Edo. 53970
Mexico
Tel: (52) 5-294-2371
Fax: (52) 2-294-8513

Rita St. Clair Associates, Inc.
Rita St. Clair, FASID
Ted L. Pearson, ASID
1009 North Charles Street
Baltimore, Maryland 21201
United States
Tel: (410) 752-1313
Fax: (410) 752-1335

Ron Wilson Designer, Inc.
1235 Tower Road
Beverly Hills, California 90210
United States
Tel: (213) 276-0666
Fax: (213) 276-7291

Charles Rose
341 Ridge Way
Carmel Valley, California 93924
United States
Tel: (408) 626-2481
Fax: (408) 626-2490

Studio Arthur de Mattos Casas
Arthur de Mattos Casas
Alameda Ministro Rocha Azevedo, 1052
São Paulo, São Paulo
Brazil 01410002
Tel: (55) 11-282-6311
Fax: (55) 11-282-6608

Studio Rublek
Aleksandar Rublek, Assoc. IALD
Johan Jongkindstraat 25
1062 CL Amsterdam
Holland
Tel: (31) 20-417-3333
Fax: (31) 20-417-3334

Angelo Tartaglia
Via Boezio, 92/D9A
00192 Rome
Italy
Tel: (39) 6-687-3879
Fax: (39) 6-686-8449

Visser & Beerman BNA
Max Zimmerman, AVB
Schouwburgplein 12-15
3055 HL Rotterdam
Holland
Tel: (31) 10-411-7814
Fax: (31) 10-414-5132

Lee von Hasseln
P.O. Box 213
Pebble Beach, California 93953
United States
Tel: (408) 625-6467

Wylie Carter Architects
J. Scott Carter
16116 Northfield Street
Pacific Palisades, California 90272
United States
Tel: (310) 459-7989
Fax: (310) 459-7292

Yarnell Associates
Bruce Yarnell, IALD, IES
12616 West 71st Street
Shawnee, Kansas 66216
United States
Tel: (913) 268-9206
Fax: (913) 268-4468

Zuhair Fayez Partnership
Zuhair H. Fayez
P.O. Box 5445
Jeddah 21422
Saudi Arabia
Tel: (966) 2-654-7171
Fax: (966) 2-654-3430

Photographers

Victor Benitez
Guanajuato 130
Col. Roma, D.F. 06700
Mexico
Tel: (52) 5-574-8032
Fax: (52) 5-584-7571

Frank Bonnet
Ransuil 5
1191 SR Ouderkerk, Amsterdam
Holland
Tel: (31) 20-496-7186
Fax: (31) 20-496-7187

Brukoff Photography
Barry Brukoff
480 Gate Five Road
Sausalito, California 94965
United States
Tel: (415) 332-6350
Fax: (415) 332-5968

Craig Mole Photography
Craig Mole
610 22nd Street
San Francisco, California 94107
United States
Tel: (415) 558-9883
Fax: (415) 626-1307

Edoardo D'Antona
Via Montesanto, 12
00195 Rome
Italy
Tel: (39) 6-560-0226

Carlos Domenech
6060 Southwest 26th Street
Miami, Florida 33155
United States
Tel/Fax: (305) 666-6964

Gil Edelstein
4120 Matisse Avenue
Woodland Hills, California 91344
United States
Tel: (818) 998-5113
Fax: (818) 998-3016

Philip H. Ennis
98 Smith Street
Freeport, New York 11520
United States
Tel: (516) 379-4273
Fax: (516) 379-1126

Eitan Feinholz
Atlaltunco 99
Tecamachalco, Edo. 53970
Mexico
Tel: (52) 5-294-2371
Fax: (52) 5-294-8513

David Glomb
458 1/2 North Genesee Avenue
Los Angeles, California 90036
United States
Tel: (213) 655-4491
Fax: (213) 651-1437

Greenworld Pictures, Inc.
Mick Hales
North Richardville Road, RD#2
Carmel, New York 10512
United States
Tel/Fax: (914) 228-0106

Philip Harvey
911 Minna Street
San Francisco, California 94103
United States
Tel: (415) 861-2188
Fax: (415) 861-2091

Julia Heine
4220 Leeward Place
Bethesda, Maryland 20816
United States
Tel: (301) 229-3705
Fax: (301) 229-6380

Lourdes Legorreta
Sierra Nevada 460
Lomas de Chapultepec 11000
Mexico
Tel: (52) 5-520-0745
Fax: (52) 5-520-4045

Ira Montgomery
2406 Converse Drive
Dallas, Texas 75207
United States
Tel: (214) 638-7288
Fax: (214) 638-7980

Mary E. Nichols
250 South Larchmont Boulevard
Los Angeles, California 90004
United States
Tel: (213) 871-0770
Fax: (213) 871-0775

Photography for the
Built Environment
Michael Spillers
421 East 69th Terrace
Kansas City, Missouri 64131
United States
Tel/Fax: (816) 444-0882

Robert Ames Cook Photography
103 Virginia Court
Franklin, Tennessee 37064
United States
Tel: (615) 591-3270

Rath & Associates
Derek Rath
4044 Moore Street
Los Angeles, California 90066
United States
Tel/Fax: (310) 305-1342

Tuca Reinés
Rua Emanuel Kant, 58
São Paulo, São Paulo
Brazil 04536050
Tel: (55) 11-306-19127
Fax: (55) 11-852-8735

Hugo Rojas
2020 North Main Street
Los Angeles, California 90031
United States
Tel: (213) 222-8836
Fax: (213) 222-5551

Samin N. Saddi
P.O. Box 5445
Jeddah 21422
Saudi Arabia
Tel: (966) 2-654-7171
Fax: (966) 2-654-3430

Douglas A. Salin
647 Joost Avenue
San Francisco, California 94127
United States
Tel/Fax: (415) 584-3322

Mark Samu
541 Lexington Avenue
New York, New York 10022
United States
Tel: (212) 754-0415
Fax: (516) 363-5242

Shinozawa Architectural
Photograph Office
Hiroshi Shinozawa
Koishikawa-Yasuda Bldg. #501
4-21-4 Koishikawa, Bunkyo-ku
Tokyo 112
Japan
Tel: (81) 3-3816-6168
Fax: (81) 3-3816-3525

Van Inwegen Photography
Bruce Van Inwegen
547 North Bernhard
Chicago, Illinois 60625
United States
Tel: (773) 583-4712
Fax: (773) 583-4812

Peter Vitale
P.O. Box 10126
Santa Fe, New Mexico 10126
United States
Tel: (505) 988-2558
Fax: (212) 838-7369

Alan Weintraub
1832A Mason Street
San Francisco, California 94133
United States
Tel: (415) 553-8191
Fax: (415) 553-8192

Ron Wilson
1235 Tower Road
Beverly Hills, California 90210
United States
Tel: (212) 276-0666
Fax: (213) 276-729

INDEX

Architects and Designers

Photographers

Acknowledgments

"Light is Life!" Motoko Ishii, the esteemed lighting designer headquartered in Tokyo once exclaimed when visiting my home in California. She was speaking, of course, about the energy and beauty of the medium, the one that brings out the spirit of all others. Beyond the magnificent fixtures she herself manufactures, beyond the wires, conduits, electric "eyes" and all her specialty's other technological aspects, it is in the end the life-enhancing effect of light that is important.

It is due to the heartfelt belief of professional designers such as Ishii in the physical, emotional and aesthetic impact of light that led to the idea for this book and to its development. Without their understanding and fervor as well as their technical knowledge and skills, this book could not have conveyed the inspiration behind the ideas that now fill its pages.

In addition, two individuals were especially helpful. Early in the planning stages, Allan Leibow, IALD, of the lighting design firm Wheel Gersztoff Shankar Selles, helped me identify many designers whose projects I included in this book. Later, on countless occasions when I was in the process of writing, architect Barry Brukoff provided invaluable advice and expertise.

As always, for this is now the ninth book we have created together, I am deeply indebted to the staff of PBC International for developing our presentation with the thoughtfulness, taste and precise execution that the subject and its practitioners deserve.

And once again I am overwhelmed by my great good fortune in knowing Angeline Vogl, my longtime associate. Even during this past year, when she has taken on the responsibilities of communications coordinator for a high school district, she continued as promised to oversee every word in this manuscript.

Finally, a most special thank you to architect Ricardo Legorreta for providing the inspiring message that is the introduction for *Designing with Light.*